Traditional Chinese Architecture Surveying and
Mapping Series:
Mausoleum Architecture & Ritual Architecture

ARCHITECTURE
COMPLEX OF SONGSHAN

Compiled by School of Architecture Tsinghua University
Edited by WANG Guixiang, LIU Chang, LIAO Huinong, HE Congrong

China Architecture & Building Press

国家出版基金项目
NATIONAL PUBLICATION FOUNDATION

『十二五』国家重点图书出版规划项目

中国古建筑测绘大系·宗教建筑与礼制建筑

嵩山建筑群

清华大学建筑学院 编写

王贵祥 刘畅 廖慧农 贺从容 主编

中国建筑工业出版社

Contents

目 录

Introduction

Since its inception in 1946, the School of Architecture at Tsinghua University has been committed to surveying and mapping traditional Chinese buildings, following the practice of the Society for the Study of Chinese Architecture (*Zhongguo Yingzao Xueshe*) that LIANG Sicheng, a driving force in the Society and founder of Tsinghua's architecture department (known as the School of Architecture since 1988), and his assistant MO Zongjiang brought with them to Tsinghua. Between 1930 and 1945, with members of the Society, LIANG visited over two thousand Chinese sites located in more than two hundred counties and fifteen provinces, and discovered, identified and mapped over two hundred groups of traditional buildings, including the famous Tang-period east hall (dating to 857) of Foguang Monastery at Mount Wutai—which was not an easy task because of the harsh working conditions in the secluded and relatively inaccessible villages in the countryside. In that same spirit, despite the difficult political circumstances from the 1950s through the 1970s, the School of Architecture conducted a systematic survey of historical buildings in the New Summer Palace (Yiheyuan). At the beginning of the Cultural Revolution in the late 1970s, all members of the faculty focusing on the history of architecture went to Hebei province under the leadership of MO Zongjiang to measure and draw the main hall of Geyuan Monastery in Laiyuan, an important Liao-period relic hidden in the remote mountains. This was followed by in-depth research and analysis. At the same time, those professors that specialized in Chinese architectural history (MO Zongjiang, XU Bo'an, LOU Qingxi, ZHANG Jingxian, and GUO Daiheng) led a group of graduate students to Zhengding in Hebei province, where they conducted component analysis and research of Moni Hall at Longxing Monastery, a Northern-Song timber-frame structure that had partially collapsed but was then in the process of being rebuilt. They also investigated nearby

导　言

因为前辈学者梁思成及其助手莫宗江两位先生从中国营造学社继承的传统，清华大学建筑学院自创立以来，一直十分注重古代建筑实例的实地考察与测绘。尽管在 20 世纪 50 至 70 年代受到各种因素的影响与冲击，那时的清华大学建筑系，还坚持了对颐和园内一批古代建筑实例的系统测绘。改革开放刚刚开始的 1970 年代末，清华大学建筑历史方向的全体教师，就在莫宗江先生带领下，共同远赴偏僻的河北山区，考察测绘了创建于辽代的涞源阁院寺大殿，并对这座辽代木构建筑进行了系统研究。同是在那一时期，建筑历史教研室的莫宗江、徐伯安、楼庆西、张静娴、郭黛姮等教师，带领研究生赴河北正定，除了对正在落架重修的北宋木构大殿隆兴寺摩尼殿的大木构件进行现场分析研究外，还对正定及周边的古建筑进行了系统考察与调研。这种由老先生带队，

historical buildings in and around Zhengding. This practice of teamwork—senior researchers, instructors, and (graduate) students participating in the investigation and mapping of traditional Chinese architecture side by side—became an academic tradition at the School of Architecture of Tsinghua University.

Since the 1980s, fieldwork has been a crucial part of undergraduate education at the School, and focus and quality of teaching has constantly improved over the past decades. In the 1990s, professors like CHEN Zhihua and LOU Qingxi carried out surveying and mapping in advance of (re)construction or land development on sites all across China that were endangered. Since the turn of the twenty-first century, the two-fold approach—attaching equal importance to practice (fieldwork) and theory (teaching)—was widened and deepened. Sites were deliberately chosen to maximize educational outcome, resulting in a broader geographical scope and spectrum of building types. In addition to expanding on the idea of vernacular architecture, special attention was paid to local (government-sponsored) construction of palaces, tombs, and temples built in the official style (*guangshi*) or on a large scale (*dashi*), and to modern architecture dating to the period between 1840 and 1949. Students and staff have accumulated a lot of experience and created high-quality drawings through this fieldwork.

In retrospect, we have completed surveys of several hundred monuments and sites built in the official dynastic styles of the Song(Jin), Yuan, Ming and Qing all across the country. Fieldwork was always combined with teaching. Among the architecture surveyed are the (single- and multi-story) buildings in front and on the sides of the Hall of Supreme Harmony in the Forbidden City in Beijing; the architecture at Changling, the mausoleum of emperor Jiaqing located at the Western Qing tombs in Yi county, Hebei province; the monasteries on Mount Wutai, Shanxi province, including Xiantongsi, Tayuansi, Luobingsi, Pusading, Nanshansi (Youguosi), and Longquansi; Zhongyue Temple, Songyang Academy, and Shaolin Monastery in Dengfeng, Henan province; Xiyue Temple, Yuquan Court, and the Taoist architecture on the peaks of Mount Hua in Weinan, Shaanxi province; Chongan Monastery, Nanjixiang Monastery, Jade Emperor Temple (Yuhuangmiao) in Shizhang, and the temples of the Two Transcendents (Erxianmiao) in Xiaohuiling and Nanshentou, all situated in Lingchuan county of Shanxi province; and the upper and lower Guangsheng monasteries and the Water God's Temple in Hongdong, Shanxi province. In recent years, we have developed a specialized interest in the study of religious architecture of Shanxi province and investigated almost a dozen privately- or government-sponsored Song and Jin sites

的学术传统。

1980 年代以来，清华大学建筑学院始终在本科教学环节中，坚持讲授古代建筑测绘这门经典课程。这一传统在 21 世纪初的这十几年中始终延续。如果说，20 世纪 90 年代由陈志华、楼庆西等教授带领的测绘教学，将相当的注意力放在了分布于全国多个省、市、自治区大量传统乡土村落建筑的抢救性测绘上，进入 21 世纪以来，清华大学建筑学院开展的这种结合本科教学的古建筑测绘教学与实践，覆盖的地域范围与建筑类型范围更为宽广：除了进一步拓展乡土建筑的测绘之外，在对各地留存的历代官式或大式建筑，如宫殿、陵寝、寺庙等建筑的测绘以及近代建筑的测绘上，也积累了大量测绘经验、图纸及丰富的调研资料。以古代官式建筑测绘为例，结合本科教学，我们先后完成了北京故宫太和殿前及两侧门殿、楼阁与朝房建筑，河北易县清西陵 昌陵 完整建筑群，山西五台山显通寺、塔院寺、罗睺寺、菩萨顶、南山寺（佑国寺）、龙泉寺等多座整组寺院建筑群，河南登封中岳庙、嵩阳书院、少林寺古建筑群，陕西渭南华山西岳庙、玉泉院及华山山顶各道观古建筑群，山西陵川崇安寺、南吉祥寺、小会岭二仙庙、南神头二仙庙、石掌玉皇庙，以及山西洪洞广胜上寺、广胜下寺、水神庙等数百座古建筑实例的测绘，其时代的范围覆盖了宋（金）、元、明、清等历代木构建筑遗存实例。近几年，我们又将测绘的重点放在了高平、晋城等晋中及晋东南地区，

located in central Shanxi (Jinzhong) and southeastern Shanxi (Jindongnan), specifically in Gaoping and Jincheng counties. This includes the Youxian, Chongming, and Kaihua monasteries and the Two Transcendents Temple in Ximen. Additionally, supported by the State Administration of Cultural Relics, the head of the Architecture History and Historic Preservation Research Institute at the School of Architecture, Liu Chang, led a group of students to map and draw the main hall of Zhenguo Monastery in Pingyao, a rare example from the Five Dynasties period. The survey results have been published. Tsinghua fieldwork in Shanxi has become an annual event that is jointly organized almost every summer by the faculty of the School of Architecture, including professors engaged in research on non-Chinese architecture, in cooperation with their graduate students.

It is worth mentioning that since 2007, the School has worked in collaboration with the well-known company China Resources Snow Breweries Ltd., which supports the transmission and dissemination of knowledge on traditional Chinese architecture and provides funds for the School's research and field investigation activities. Drawing on the support from industry allowed us greater initiative and flexibility, and we were thus able to carry out research on and survey often overlooked but no less important Song-Jin monuments in central and southeastern Shanxi.

Our years-long fieldwork has not only enabled us to teach students subject knowledge about scale, material, form, and decoration of traditional Chinese architecture as well as a sense of appreciation for the old, but has also provided us with plenty of data for monument preservation practice and research. China Architecture and Building Press spared no effort in compiling and publishing the results of the fieldwork in 2012. Publication has also been supported by the National Publishing Fund. This highlights not only the importance of our contribution to architectural education at the national level but also shows its significance for the transmission, development, and revival of traditional Chinese architectural culture both at home and abroad. In order to expand the reach of this work to an international audience, *the Traditional Chinese Architecture Surveying and Mapping Series* is being published bilingually. Based on the past ten years of fieldwork, we have now compiled five volumes, namely *Mount Wutai's Buddhist Architecture* (Traditional architecture on Mount Wutai, Shanxi), *Architecture Complex of Songshan* (Traditional architecture in Dengfeng, Henan), *Mount Hua's Yuemiao and Taoist Temples* (Traditional architecture on Mount Hua, Shaanxi), *Architecture Complex of Hongtong* (Traditional architecture in Hongtong, Shanxi), and *Architecture Complex*

对包括高平游仙寺、崇明寺、开化寺、西李门二仙庙等在内的十余座宋金建筑群，进行了全面而系统的测绘。这一期间，在国家文物局的支持下，建筑历史与文物保护研究所刘畅老师还带领研究生对五代时期创建的平遥镇国寺大殿等建筑进行了精细测绘，并出版了测绘研究成果。此外，清华大学建筑学院的测绘工作，几乎每年都是由全体建筑历史教师共同合作，并带领研究生们共同完成的。从事外国建筑史教学的老师，也不例外。

特别值得一提的是，自 2007 年以来，清华大学建筑学院与国家知名企业华润雪花啤酒（中国）有限公司建立了良好的合作关系。该集团不仅支持中国古建筑知识的传承与普及工作，也对清华大学建筑学院中国古代建筑研究及古建筑测绘工作给予了直接的支持，使得我们的古建筑测绘工作变得更为主动和更具选择性。一大批珍贵的山西晋中及晋东南地区宋金时代建筑实例的测绘与研究，就是在这样一个前提下得以顺利开展与完成的。

坚持数十年的古建筑测绘工作，不仅在培养学生对传统中国建筑的尺度、材料、造型与细部装饰的认知与感觉上起到了直接的影响，而且也为各地文物建筑保护与研究工作，提供了相当充分的资料支持。

2012 年，中国工业出版社花大气力组织了汇集全国重点院校建筑系古建筑测绘成果的中国古代建筑测绘大系的编辑出版工作。这一工作也获得了国家出版基金的支持。这不仅是对高校建筑教育成果的一份支持，也是对中国传统建筑文化传承、发展与复兴的一份支持。正是在这样一个背景与前提下，我们对近十余年来考察测绘的古代建筑案例加以整理，分别编汇了包括《五台山佛教建筑》《嵩山建筑群》《华山岳庙与道观》《洪洞建筑群》《高平建筑群》5 册古建筑测

of Gaoping (Traditional architecture in Gaoping, Shanxi). The architectural drawings presented in these books are carefully selected and screened by Tsinghua professors. They only show a part of our comprehensive surveying and mapping work, but still cover a whole spectrum of geographic regions and time periods. Thus, they contain information of high academic value that may serve as a reference for future study and for the protection of cultural heritage. It is hoped that our work will help to promote interest in and improve understanding of traditional Chinese architecture, not only among Tsinghua students (through hands-on experiences in the fieldwork) but also among architectural historians and professionals engaged in monument preservation at home and abroad.

As a final thought, let me shortly address the workflow. The drawings presented here are based on survey and working sketches drawn up on site during several years of fieldwork conducted by Tsinghua professors together with graduate and undergraduate students. Back home, the measured drawings were redrawn over months of diligent work by graduate students with computer-aided software to achieve dimensionally accurate and visually appealing results, a project that was completed under the supervision of LIU Chang, head of Tsinghua's Architecture History Institute, and the Tsinghua professors LIAO Huinong and WANG Nan, as well as TANG Henglu and his colleagues from the WANG Guixiang Studio. We would like to take this opportunity to thank the professors, students and colleagues who participated in the fieldwork and its revision.

Our final thanks go to LI Jing, assistant researcher at the Architecture History Institute here at Tsinghua. Next to participating in surveying and mapping, she organized the development of the book and moreover, made this book possible in the first place.

WANG Guixiang, LIU Chang, LIAO Huinong
Architecture History and Historic Preservation Research Institute, School of Architecture,
Tsinghua University
December 5, 2017

Translated by Alexandra Harrer

绘图集，作为这套『中国古建筑测绘大系』的部分成果。尽管这只是我们多年测绘成果的一部分，

但也是清华建筑历史学科教师们仔细筛选、认真校对、充分整理之后的较具典型性与参考性的成果。

这些成果不仅地域覆盖面大，而且建筑遗存的时代跨度也相当长，具有十分重要的学术价值。希望

这些成果对高校建筑系学生们学习古建筑，建筑历史学者研究古建筑，以及文物保护工作者从事文

物古建筑的保护与修缮，能够起到积极的推动作用与重要的参考价值。

最后要提到的一点是，除了参与测绘的教师、研究生与本科生多年历尽辛苦的测量与绘图工作

之外，此次清华大学建筑学院承担的这 5 册测绘图集，也经由建筑历史与文物保护研究所刘畅、廖

慧农、王南和他们的研究生，以及王贵祥工作室团队的唐恒鲁等同仁们在既有测绘图纸基础上，经

过数月认真仔细的线条分层、图面调整、数据校对、图面完善等缜密修复工作，在这里也要向参加

测绘图整理的老师、同学和同事们表示感谢。

还应该特别提到的是建筑学院建筑历史与文物保护研究所的助理研究员李菁博士，她不仅参加

了多次测绘，还为这套书最后的编辑与出版做了大量相关工作。这里一并表示感谢。

清华大学建筑学院 建筑历史与文物保护研究所

王贵祥、刘畅、廖慧农

2017 年 12 月 5 日

Preface

I.Significance and Purpose of Surveying and Mapping the Historic Monuments on Mount Song

Songshan or Mount Song is located in the hinterland of the Central Plains northwest of the county-level city of Dengfeng in western Henan province. After king Ping of Zhou moved the capital to nearby Luoyang in 770 BCE, Songshan became "the center of heaven and earth"—the only point where astronomical observations were considered to be accurate. This was based on an ancient saying that established the geographical distribution of the Five Great Mountains in China ("Mount Song occupies the center, to the left is Mount Tai, to the right is Mount Hua"). As the "Holy Peak of the Center" or more broadly, the Central Mountain, Mount Song became the subject of imperial pilgrimage throughout the centuries with emperors holding a grand ceremony of worship of heaven on top of the mountain and praying for peace and prosperity. Songshan is also key to understanding the "Three Teachings"—Confucianism, Buddhism, and Taoism—because it played a key role in the formation and cultural propagation of these three religious and philosophical traditions. The historical buildings situated in and around Dengfeng are the living testimony of this eventful history. Among them are the Taishi Gate-towers (*que*), Zhongyue Temple, Shaoshi Gate-towers (*que*), Qimu Gate-towers (*que*), the pagoda of Songyue Monastery, the Shaolin Monastery cluster (namely the Permanent Residence Court [Changzhuyuan], Hermitage of the First Patriarch [Chuzu'an], and Pagoda Forest [Talin]), Huishan Monastery, Songyang Academy, and an astronomical observatory. These buildings have a centuries-long history (from the Han to the Qing) and comprise a variety of different types (ritual architecture, religious architecture, and science and education architecture). As a whole, they constitute a group of cultural monuments with outstanding universal value (Figs.1, 2).

序言

一、嵩山古建筑测绘的意义与目的

嵩山位于中原腹地，今河南省西部登封市的西北部。自周平王迁都洛阳，以『嵩为中央、左岱、右华』，嵩山领『天地之中』，即成为历代帝王祭祀封禅的对象，以及三教策源、文化荟萃的场所，继而为今天留下了一系列重要的历史建筑遗存，包括太室阙和中岳庙、少室阙、启母阙、嵩岳寺塔、少林寺建筑群（常住院、初祖庵、塔林）、会善寺、嵩阳书院、观星台等，分布在登封市区周围。这些建筑群时间跨度长（从汉到清）、建筑种类丰富（覆盖宗教建筑、礼制建筑等），是一批极其珍贵的建筑文化遗产（图1、图2）。

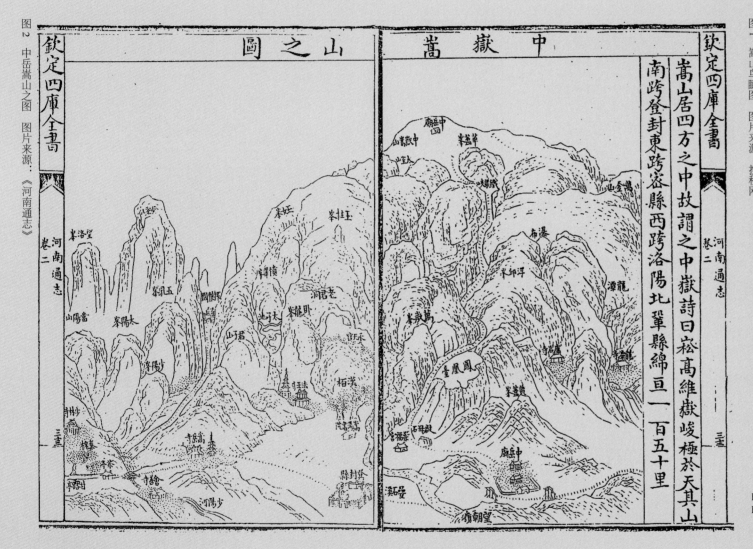

图一　嵩山鸟瞰图　图片来源：携程网

图2　中岳嵩山之图　图片来源：《河南通志》

Fig.1 Aerial view of Songshan (Mount Song) Source: Website of Ctrip

Fig.2 Map of Songshan (Mount Song), the central mountain among five great mountains
Source: *Local Records of Henan Province*

Surveying and mapping of historical architecture is an effective means of monument preservation and of documentation of the current condition of a building. Each field survey establishes a reference framework and collects information about a building otherwise lost, providing a reliable basis for future conservation and repair. At the same time, each field survey provides raw data for a more comprehensive and systematic study of the historical architecture on Songshan. It similarly provides specific information about the regional building style in the Songshan area (Fig.3).

In terms of architectural education and professional training, first-hand experience in dealing with scale, proportion, construction, and details of historical architecture is necessary for a rational and emotional understanding of the true significance of one's own cultural heritage. The engagement of students throughout the entirety of this survey project — ranging from measuring and drawing by hand to converting the collected data into computer-generated technical drawings — is a rare opportunity for them to learn. The various experiences and unforgettable memories that teachers and students take with them from their field trips have also enhanced the dialog between academics (teachers and students) and the professionals in charge of the Songshan monuments.

In July 2007, the Municipal Cultural Bureau of Zhengzhou, Henan province, invited the School of architecture of Tsinghua University to assist in the preparation of the nomination proposal for inclusion of Songshan in the World Heritage List. A team of ninety-eight teachers and students, led by the Tsinghua professors WANG Guixiang, LIAO Huinong, LIU Chang, and HE Congrong, took measurements of three groups of buildings —Zhongyue Temple, Changzhu Court at Shaolin Monastery, and Songyang Academy. The aim of surveying and mapping was to bring to light the significance of Songshan for the cultural history of rituals, religions, and science and education.

The following paragraphs introduce the three groups of buildings that we surveyed.

将这批珍贵的建筑文化遗产，通过建筑测绘的方式保存下原始档案，是建筑文物保护工作者们的一项工作必需。每测绘一座古建筑，就为这座建筑建立了一个档案，从而为这座建筑的保护、研究，甚至一旦遭到破坏时的正确修缮，提供了一个可靠的依据。同时，也为嵩山古建筑的深入研究提供一点原始资料，甚至为嵩山地区古建筑保护与传统风格建筑的创作与传承提供一些真实而具体的案例（图3）。

对于建筑教学，通过建筑测绘全过程，可以令建筑系师生与嵩山古建筑实地踏勘数日，亲手触摸这些经典的古代建筑作品，亲身体味其尺度、比例、构造与细部，亲手获取建筑数据并绘制出矢量化图形，从而完成对古建筑经典从直观感受到建构认知，从草图勾勒到数字化表达的深化，是非常难得的学习机会。对此一生一次的古建测绘经历，也会成为本科生们毕生难忘的回忆。这批建筑界将来的骨干，与珍贵的古建筑遗产抑或以此结缘。

2007年7月，应河南郑州市文化局的邀请，为了配合嵩山古建筑群申报世界文化遗产的工作，清华大学建筑学院师生一行98人，由王贵祥、廖慧农、刘畅、贺从容等老师带队，分三组分别对中岳庙、少林寺常住院和嵩阳书院三组建筑群进行了测绘，测绘目标涵盖了嵩山在宗教文化、礼制文化及科教文化方面的古建遗存。

接下来，简要介绍一下收入本册测绘图集的三组重要古建筑群。

图3 嵩山古建筑分布图 图片来源：扫描自《嵩山申遗文本》

Fig.3 Location map of traditional Chinese architectures in Songshan (Mount Song) Source: *Documents of Nominating Songshan as World Cultural Heritage*

II. Introduction to the Historic Monuments on Mount Song

(1) Shaolin Monastery

Set against the backdrop of Wuru Summit, Shaolin (Shao[shi] Mountain Forest) Monastery is oriented southward toward Shaoshi Mountain, a location from which it derives its name. A line in *Shuo Song* by the Qing scholar Jing Rizhen explains this, saying: "Shaolin means forests of Shaoshi (Mountain, one of the mountains that make up Mount Song)".

In 495 (the nineteenth year of the Taihe reign period of the Northern Wei dynasty), emperor Xiaowen built a temple for the Dhyana master Batuo to settle down and preach. In 527 (the third year of the Xiaochang reign period of the Northern Wei dynasty), the eminent Indian monk Bodhidharma, first Chinese patriarch and transmitter of Chan Buddhism to China and, became abbot of Shaolin Monastery. At the end of the Sui, armed peasants razed the temple. Because Shaolin monks rendered great service to Li Shimin (in one of his military campaigns to reunify the empire), they were rewarded with imperial patronage after his accession to the throne as (the second) Tang emperor Taizong. The monastery was renovated on several occasions during the Tang period, the golden age of the monastery, resulting in an architecture and scale that was magnificent by the time of emperor Xuanzong. In the Great Anti-Buddhist Persecution under emperor Wuzong, many buildings were destroyed. They were rebuilt in the Yuan dynasty but at the end of that period, the monastery was ransacked by soldiers and endangered again. It was finally restored in the Ming dynasty to its present-day scale, and a sutra pavilion, Lixue (Standing-in-snow) Pavilion, and Qianfo (Thousand Buddha) Hall were added at that time (Figs.4~11).

The complex has an area of 3.6 hm^2, measuring 300 m in the north-south direction and 120 m in the east-west direction. The principal structures include the following monuments: Changzhu (Permanent Residence) Court; to the west of it, the Pagoda Forest; to the north, the Hermitage of the First Patriarch, Bodhidharma Cave, and Ganlu Platform; to the southwest, the Hermitage of the Second Patriarch; and to the northeast, the Hermitage of Guang Hui. Scattered around the monastery are more than ten Chan pagodas such as those of the masters Tong Guang, Fa Ru, and Fa Hua.

Changzhu Court has seven courtyards arranged one after the other. The main structures

二、嵩山古建筑群的简要介绍

（一）少林寺

『少林者，少室之林也。』（清景日珍《说嵩》）寺院坐北朝南，背靠五乳峰，面对少室山，少林寺因此得名。北魏太和十九年（495年），孝文帝元宏为安顿天竺僧人跋陀敕建少林寺于嵩山。北魏孝昌三年（527年）印度高僧菩提达摩主持少林寺，在这里首传禅宗，所以被尊为禅宗祖庭。少林寺隋末曾遭农民军火焚，但因少林寺武僧助战李唐有功，得到唐太宗封赏。唐时几经修整，寺院发展到鼎盛，至玄宗年间，建筑规模已相当宏伟。唐武宗灭法期间，不少殿宇被毁，元代有所兴建。元末兵火后，明代又重修了藏经阁、千佛殿、立雪亭等，奠定了今日少林寺之规模（图4～图11）。

寺院坐北朝南，南北长300米，东西宽120米，占地面积3.6公顷。主要包括常住院，寺西有塔林，

图四　少林寺鸟瞰图　图片来源：扫描自《嵩山申遗文本》

Fig.4　Aerial view of Shaolin Monastery
　　　Source: *Documents of Nominating Songshan as World Cultural Heritage*
Fig.5　Archway at entrance of Shaolin scenic spot
　　　Source: Photographed by HE Congrong
Fig.6　Scenery of the entrance, Shaolin Monastery
　　　Source: Photographed by HE Congrong

图六　少林寺入口外景　图片来源：贺从容摄

图五　少林寺景区入口牌坊　图片来源：贺从容摄

are aligned along a central axis, comprising a mountain gate, Hall of Heavenly Kings (Tianwangdian), Treasure Hall of the Great Hero (Daxiongbaodian), a sutra pavilion, an abbot's hall, Lixue (Standing-in-snow) Pavilion, and Qianfo (Thousand Buddha) Hall. The ground rises slightly to the north, resulting in a 22.61-m difference in the terrain. Buildings from the Ming-Qing period— whether they are topped with a gable-roofed, hip-roofed, or hip-gable roofed—usually have one set of eaves covered with glazed tiles and are of moderate volume. The only exception is Thousand Buddha Hall, a seven-bay structure embellished with precious murals painted on a surface of more than 300 m sq. Next to the central row of buildings stand a bell tower, a drum tower, Liuzu (Six Patriarchs) Hall, Jinnaluo (Kinnara) Hall, east- and west-side buildings (*liaofang*), Wenshu (Manjusri) Hall, Puxian (Samantabhadra) Hall, Baiyi (White Robe) Hall, Dizang (Ksitigarbha) Hall, and several auxiliary buildings. The Pagoda Forest is situated to the west of Changzhu Court and to the east are meditation halls (*chantang*) and monks' rooms (*sengfang*). Additionally, two Song-period pagodas dedicated to Shakyamuni Buddha and to the future Buddha Maitreya, one hundred and seventy-four stone steles from successive dynasties, and several 1500-year old Ginko trees are also preserved in Changzhu Court (Figs.12~18).

北有初祖庵、达摩洞、甘露台，西南有二祖庵，东北有广慧庵。寺周还有同光禅师塔、法如禅师塔和法华禅师塔等古塔 10 余座。少林寺常住院中轴线有山门、天王殿、大雄宝殿、藏经阁、方丈室、立雪亭、千佛殿等七进建筑，阶梯升高，高差 22.61 米。其明清建筑多为单檐琉璃瓦，硬山、歇山、庑殿间而有之，除千佛殿外，体量适中。千佛殿面阔七间，殿内墙壁满绘壁画达 300 多平方米，甚为珍贵。中轴线两侧有钟楼、鼓楼、六祖殿、紧那罗殿、东西寮房、文殊殿、普贤殿、白衣殿、地藏殿等附属建筑。再往外，西有塔院，东有禅堂、僧房等，常住院内还保有宋代释迦塔、下生弥勒佛塔，历代碑碣 174 品以及秦槐、树龄 1500 多年以上的银杏等古树名木（图 12～图 18）。

图7 少林寺山门前牌坊 图片来源：贺从容摄

图8 少林寺山门 图片来源：廖慧农摄

图9 少林寺立雪亭 图片来源：贺从容摄

图10 少林寺大雄宝殿 图片来源：廖慧农摄

Fig.7 Archway at mountain gate, Shaolin Monastery
Source: Photographed by HE Congrong
Fig.8 Mountain gate of Shaolin Monastery
Source: Photographed by LIAO Huinong
Fig.9 Lixue Pavilion of Shaolin Monastery
Source: Photographed by HE Congrong
Fig.10 Treasure Hall of Great Hero, Shaolin Monastery
Source: Photographed by LIAO Huinong

图12　少林寺白衣殿　图片来源：贺从容摄

图11　少林寺千佛阁　图片来源：廖慧农摄

图14　少林寺千佛殿　图片来源：贺从容摄

图13　少林寺白衣殿室内　图片来源：贺从容摄

Fig.11　Qianfo Pavilion of Shaolin Monastery　Source: Photographed by LIAO Huinong
Fig.12　Baiyi Hall of Shaolin Monastery　Source: Photographed by HE Congrong
Fig.13　The indoor scene of Baiyi Hall, Shaolin Monastery　Source: Photographed by HE Congrong
Fig.14　Qianfo Hall of Shaolin Monastery　Source: Photographed by HE Congrong

图15 少林寺方丈室 图片来源：贺从容摄

图16 少林寺禅堂斋堂 图片来源：贺从容摄

图17 少林寺山门内甬道 图片来源：贺从容摄

图18 少林寺藏经阁后边门 图片来源：贺从容摄

Fig.15 Abbot's Hall of Shaolin Monastery
　　　Source: Photographed by HE Congrong
Fig.16 Meditation hall and abstinence hall of Shaolin Monastery
　　　Source: Photographed by HE Congrong
Fig.17 Path inside the mountain gate of Shaolin Monastery
　　　Source: Photographed by HE Congrong
Fig.18 Side door of sutra pavilion, Shaolin Monastery
　　　Source: Photographed by HE Congrong

(2) Songyang Academy

Songyang (South [of Mount] Song) Academy is situated below Junji Summit on the southern slope of Taishi Mountain and named after its exposed location on Mount Song—surrounded by mountains on the east, north, and west sides, only the terrain in the south is flat and reached by sunlight (*yang*). The academy was formerly known as Songyang Monastery, founded in 484 (the eighth year of the Taihe reign period of the Northern-Wei emperor Xiaowen) and at one point, it housed several hundred monks. During the Daye reign (605-618) of the Sui emperor Yang, the complex was turned into a Taoist temple. Repaired in the Five Dynasties and modified into an academy for classical learning, it was only in 1035 (the second year of the Jingyou reign period of the Song emperor Renzong) that the complex was named Songyang Academy. Since then, famous priests and celebrities have visited the site and given lectures on Chinese classics.

Songyang Academy was the center of Song and Ming Neo-Confucianism. Cheng Yi and Cheng Hao, founders of the Luo School of Song Neo-Confucianism, were renowned scholar-officials whose thoughts influenced society even in late imperial China. They taught at Songyang Academy. During the military campaigns at the end of the Ming dynasty, Songyang Academy was destroyed by fire but rebuilt shortly afterwards in the Qing dynasty. During the reign of the Qing emperor Kangxi, the famous Confucian scholar Geng Jie used his private financial resources to restore and expand the academy into a center for the teaching of the Luo School of Neo-Confucianism. Songyang Academy occupies an important position in Chinese cultural history as part of the Confucian academic system of higher education. Together with Suiyang Academy in Shangqiu, Henan province, Yuelu Academy in Changsha, Hunan province, and Bailudong Academy on Mount Lun in Jiangxi province, it was one of the four great imperial academies. In 1998, the State Post Bureau issued a special series of stamps of the four academies.

（二）嵩阳书院

嵩阳书院位于太室山南麓，峻极峰下，东、北、西三面山峰环抱，南面开阔平缓，因坐落在山之阳而得名。书院前身为嵩阳寺，创建于北魏孝文帝太和八年（484 年），僧众曾多达数百人。隋炀帝大业年间（605—618 年）改成道教活动场所，更名为嵩阳观。五代后周时改建为书院，宋仁宗景祐二年（1035 年）名嵩阳书院，此后一直是历代名人讲授经典的教育场所。

嵩阳书院是宋明理学教育中心之一。程颐、程颢创建的宋明理学，而嵩阳书院又是『二程』等著名儒学大家活动过的地方，所以历代官吏更名儒对嵩阳书院都曾有修复。明末，嵩阳书院毁于兵火，清各代重修增建，清代康熙年间，名儒耿介倾资大规模修复并扩建了嵩阳书院，使它成为清代洛派理学的传播中心。嵩阳书院在中国文化史中占有重要地位，与河南『睢阳书院』、湖南『岳麓书院』、江西『白鹿洞书院』并称为宋初四大书院。

1998 年，国家邮政局还特别发行了上述古代四大书院的邮票。

图19
嵩阳书院鸟瞰图
图片来源：《嵩山申遗文本》

图20
嵩阳书院山门
图片来源：廖慧农摄

图22
嵩阳书院藏书楼
图片来源：贺从容摄

图21
嵩阳书院讲堂
图片来源：廖慧农摄

图24
嵩阳书院先圣殿
图片来源：贺从容摄

图23
嵩阳书院道统祠
图片来源：贺从容摄

Fig.19　Aerial view of Songyang Academy
　　　　Source: *Documents of Nominating Songshan as World Cultural Heritage*
Fig.20　Mountain gate of Songyang Academy　Source: Photographed by LIAO Huinong
Fig.21　Lecture hall of Songyang Academy　Source: Photographed by LIAO Huinong
Fig.22　Library of Songyang Academy　Source: Photographed by HE Congrong
Fig.23　Daotong Shrine of Songyang Academy　Source: Photographed by HE Congrong
Fig.24　Xiansheng Hall of Songyang Academy　Source: Photographed by HE Congrong

The south-facing architecture of Songyang Academy has largely retained its Qing-dynasty layout and covers an area of 9.984 m sq, measuring 128 m in the north-south direction and 78 m in the east-west direction. Five courtyards are arranged along the central axis and, from south to north, comprise a principal gate, Xiansheng Hall, a lecture hall, Daotong Shrine, and a library. East- and west-side halls (*peifang*) stand on both sides of the central axis. Buildings are designed in the local style of Henan province. They have a flush gable roof with rounded ridge covered with glazed tiles (except from the hip-gable roofed Daotong Shrine) and are elegant and graceful in appearance. Twenty-six buildings from the Qing dynasty have survived, next to fifteen stone carvings from the Eastern Wei dynasty and fourteen ancient trees. Songyang Academy, with its well-distributed and functional layout, is key not only to the understanding of Confucian culture in imperial China but also to the study of traditional Chinese academies and buildings of higher learning. Even emperor Qianlong praised the enchanting and inspiring atmosphere of the academy and its architecture in his poetry (Figs.19~24).

As a typical example of traditional Chinese ritual architecture, religious architecture, and educational architecture, Zhongyue Temple, Shaolin Monastery, and Songyang Academy were inscribed on the World Heritage List as part of the "Historic Monuments of Dengfeng in 'The Centre of Heaven and Earth'".

嵩阳书院建筑群坐北朝南，基本保持了清代的建筑布局，南北长128米，东西宽78米，占地面积9984平方米。沿中轴线布置有五进院落，由南向北依次为大门、先圣殿、讲堂、道统祠和藏书楼，中轴线两侧有相连的配房。除道统祠为歇山顶以外，其他建筑均为硬山卷棚布瓦顶，显得典雅大方，具有河南地方建筑风格。书院现存有清代建筑26座，还保存了东魏以降的石刻15品、古树14株等历史遗存。嵩阳书院格局紧凑，功能完善，对研究我国古代书院建筑、教育制度以及儒家文化具有不可替代的标本意义。乾隆曾有诗云："书院嵩高景最清，石幢犹记故宫名。山色溪声留宿雨，菊香竹韵喜新晴。初来岂得无言别，汉柏阴中句偶成。"（图19～图24）

作为中国古代礼制建筑、宗教建筑、科教建筑的典型代表，作为"登封「天地之中」历史建筑群"的子项目，中岳庙、少林寺、嵩阳书院均已被联合国教科文组织正式列入世界文化遗产名录。

图 25　庙会图　图片来源：《嵩岳庙史》景日畛撰（清康熙）

Fig.25 Drawing of temple fair Source: *The History of Songyue Temple* by JING Rizhen（In Kangxi reign, Qing Dynasty）

(3) Zhongyue Temple

Zhongyuemiao, or Temple to the Central Peak (referring to Songshan, the Central Mountain of the Five Great Mountains), is located on Huanggai Summit on the southern slope of Taishi Mountain—one of the mountains that make up Mount Song—three kilometers east of Dengfeng. In spite of repair on several occasions, the ancestral temple system has been preserved to this day. The first structure built at the site (in the Qing dynasty) was known as Taishi Shrine and was a place where the local gods and spirits of Taishi Mountain were worshiped. After repair under the Han emperor Wu, the shrine was renamed Zhongyue Temple in the Northern Wei dynasty. It was substantially enlarged on several occasions in the Tang-Song period. With the disappearance of systematic ritual offering to sacred mountains, the temple became a place for Taoist activities. Famous Taoist masters wrote books here and held lectures. It became known as the Sixth Lesser Grotto-heaven (Diliu xiaodongtian), where, according to legend, Prince Jin of Zhou had become a Taoist immortal. The temple was destroyed by fire in the seventeenth year of the Chongzhen reign period of the Ming dynasty, and subsequently repaired in the Qing dynasty. Today, Zhongyue Temple belongs to the Quanzhen School of Taoism and has largely retained the grand scale of the Ming-Qing period, showing formal and stylistic characteristics of late imperial government-sponsored construction (Figs.25~28).

As a venue for official state sacrifices in addition to being a sacred site (grotto-heaven) for Taoist devotion, Zhongyue Temple has been highly valued since ancient times for its close adherence to the Chinese system of ritual as well as codified building regulation. Thus, it is not surprising that Zhongyue Temple is an architectural complex of extremely large scale with an area of more than 10.0000 m sq, measuring 650 m the in the north-south direction and 166 m in the east-west direction. The temple architecture fits well into the mountainous landscape—the terrain rises gradually towards the north resulting in a north-south elevation difference of 37 m. Located to its south at a distance of 600 m were the Taishi Gate-towers (*que*). With these markers of a ceremonial gateway in the front and a towering (palace-)city wall with four gates all around, the temple came to resemble a small town both practically—because of its inherently enhanced defenses—and conceptually—the imitation of imperial palace(-city) construction increased the significance of worshipping the mountain gods and spirits.

Zhongyue Temple is oriented southward, and the beauty of its architecture unfolds in a series of linked courtyards that vary in size and design. Eleven buildings are aligned along

（三）中岳庙

中岳庙位于登封市区东3公里太室山南麓的黄盖峰下，历经变迁增修，乃至庙制完善，遗存至今。

中岳庙的前身叫作『太室祠』，是祭祀嵩山太室山神的场所。汉武帝时曾有增修，北魏时改名为中岳庙。唐宋时多次扩建，规模渐大。随着祭祀山岳制度的消失，中岳庙演变成道教的活动场所，历代名道士曾在此著书讲经，道家尊其为『道教第六小洞天』，认为这里是周朝的神仙王子晋的升仙之处，中岳庙成为道教主流全真道的圣地。明崇祯十七年中岳庙毁于大火，清代多次重修，今天的中岳庙基本上保留了当时的宏伟规模，具有明清官式建筑的规模格局和风格特点（图25～图28）。

作为国家官方祭祀和道教洞天福地的场所，中岳庙自古以来备受隆崇，其礼制等级和建筑规制都很高。整组祠庙占地十万多平方米，南北全长650米，东西宽166米，规模宏大壮观。建筑群依山势而建，自南向北渐次升高，前后高差37米，颇有气势。庙门南600米有太室阙，庙四周有庙墙高筑、开设四门，规制之高无异于一座小城。如此做法，一是模拟帝王宫室以敬岳神，二是重垣高墙起到防卫的作用。

中岳庙坐北朝南，依南北中轴线左右对称，以纵深多进、多群组院落的形式组合而成。沿中轴线存有十一进建筑，依次为中华门、遥参亭、天中阁、配天作镇坊、崇圣门、化三门、峻极门、崇

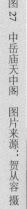

图27 中岳庙天中阁 图片来源：贺从容摄

图28 中岳庙崇圣门 图片来源：贺从容摄

图26 中岳庙鸟瞰图 图片来源：扫描自《嵩山申遗文本》

Fig.26 Aerial view of Zhongyue Temple Source: *Documents of Nominating Songshan as World Cultural Heritage*
Fig.27 Tianzhong Pavilion of Zhongyue Temple Source: Photographed by HE Congrong
Fig.28 Chongsheng Gate of Zhongyue Temple Source: Photographed by HE Congrong

图 29　中岳庙中轴节点　图片来源：贺从容摄

Fig.29　The joint of north—south axis　Source: Photographed by HE Congrong

图 31　中岳庙中岳大殿　图片来源：贺从容摄

图 30　中岳庙峻极门　图片来源：廖慧农摄

Fig.30 Junji Gate of Zhongyue Temple Source: Photographed by LIAO Huinong
Fig.31 Zhongyue Hall of Zhongyue Temple Source: Photographed by HE Congrong

the north-south axis and arranged symmetrically in relation to it—Zhonghua Gate, Yaocan Pavilion (*ting*), Tianzhong Pavilion (*ge*), Peitian zuozhen Archway (*fang*), Chongsheng Gate, Huasan Gate, Junji Gate, Chonggao junji Archway (*fang*), Zhongyue Hall (also known as Junji Hall), a resting hall (*qindian*), and an imperial library (Figs.29~31). Several buildings still stand on the side of the central axis, including a hexagonal pavilion (*ting*), the platform of Siyue (Four Peak) Hall, the imperial stele pavilion and corridor buildings of the front courtyard, the corridor buildings of the back courtyard, and east- and west-side buildings attached to both sides of the imperial library.

Among them, Zhongyue Hall is the largest and most magnificent building of the temple. Measuring nine-by-five bays it is embellished with decoratively carved window and door grills. The double-eaves hip-gable roof is covered with yellow glazed tiles. Eave bracket sets are of seventh or ninth rank (Qing carpentry). An artistically carved caisson ceiling (*zaojing*) with a dragon in the center, a feature used only in high-rank architecture, is installed in the center of the hall and surrounded by flat, coffered ceilings (*tianhua*). The hall was repaired in the sixth year of the Dazhong xiangfu reign period of the Song dynasty, then destroyed by fire in 1641 (the fourteenth year of the Chongzhen reign period of the Ming dynasty), and rebuilt again in 1653 (the tenth year of the Shunzhi reign period of the Qing dynasty).

The small gate (*yemen*) adjacent to the eastern side of Junji Gate contains a stone stele (Da Jin Chengan zhongxiu Zhongyuemiao tu) (Fig.32) that shows, in a traditional flat-axonometric projection ("elevation-on-plan"), how the temple might have looked seven hundred and eighty years ago.

Several independent courtyards are arranged to the east and west of the central courtyard row—Taiwei Palace, Huoshen Palace, Zushi Palace, Xiaolou Palace, and Longwang Hall As a whole, the architecture of Zhongyue Temple comprises thirty-nine buildings (halls [*dian*], palaces [*gong*], multi-story buildings [*lou*], multi-story pavilions [*ge*], single-story pavilions [*ting*], and corridors [*lang*]) and adds up to four hundred bay-lengths. The temple layout can be traced back to the Jin dynasty, but the individual buildings were rebuilt on several occasions during the Qing dynasty. In addition, there are seventy-three stone steles, three hundred and thirty cypress trees planted from the Han to the Qing, and several bronze vessels (*ding*) and figures (*tieren*). Emperor Qianlong once visited Zhongyue Temple, and during his stay, he composed two appreciation poems (entitled *Yeyuemiao*).

高峻极坊、中岳大殿（峻极殿）、寝殿、御书楼等主要建筑（图29～图31）。中轴线的两侧，还保存有六角亭、四岳殿台基、大殿院的御碑亭和廊庑、寝殿院的廊房以及御书楼两端的东西顺山房等建筑。其中，核心建筑中岳大殿（峻极殿）是庙中规模最大的殿宇，面阔九间，进深五间，高大雄伟、十分壮观；重檐庑殿顶，黄琉璃瓦，上下檐分别施七踩和九踩斗栱，透花棂子门窗。殿内中部有精雕的盘龙藻井，满饰天花，等级很高。大殿建筑增修于宋真宗大中祥符六年（1013年），明崇祯十四年（1641年）大殿毁于火，清顺治十年（1653年）重建。另外，峻极门、东掖门内有『大金承安重修中岳庙图』碑（图32），用传统的主体透视刻绘，保存了780多年前中岳庙的全貌。

庙的东西两路，分别建有太尉宫、火神宫、祖师宫、小楼宫和龙王殿等独立小院。庙中共有殿、宫、楼、阁、亭、廊等建筑39座近400间。现存建筑格局可上溯到金代，建筑多复建于清代。另外，庙内还有古代碑刻73品，有汉到清代古柏330余株，及神鼎、铁人等文物小品，也是难得的历史景观。

清乾隆皇帝曾至中岳庙致祭，当夜御制《谒岳庙》诗二首，其一为：『正正堂堂地，巍巍焕焕京。到来瞻气象，果足庆平生。惬我长年愿，陈兹祈岁情。忽闻鸾鹤韵，疑有列仙迎。』

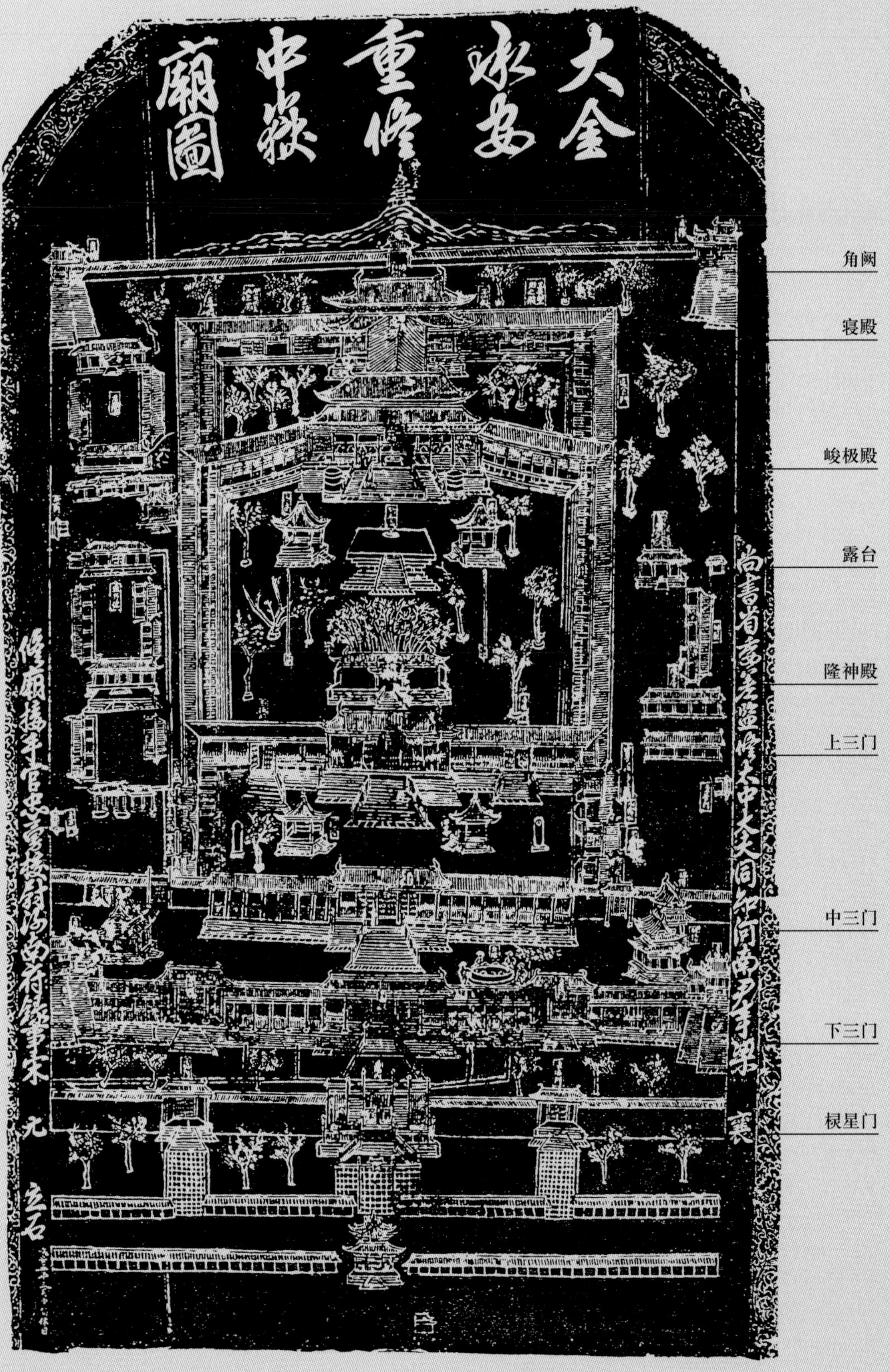
大金承安重修中嶽廟圖
角阙
寝殿
峻极殿
露台
隆神殿
上三门
中三门
下三门
棂星门

图32 大金承安重修中岳庙图
图片来源：傅熹年《中国古代城市规划、建筑群布局及建筑设计方法研究》

Fig.32 Reconstruction of Zhongyue Temple in Cheng'an Period, Jin Dynasty Source: *Research on Traditional Chinese City Planning, Layout of Architectural Complex and Architectural Design Methodology* by FU Xinian

025

Zhongyue Temple has a clear hierarchical structure—important buildings of high rank can be easily distinguished from ordinary buildings—and a well proportioned, well distributed, and harmoniously balanced layout. It is the largest and best-preserved site among all the temples to any of China's famous mountains or peaks (wuyue), and the largest and most complete architectural ensemble in Henan province. Zhongyue Temple is a perfect example of traditional Chinese ritual architecture.

III. Introduction to the Surveying and Editing Work

The fieldwork program lasted two weeks. In the first week, we completed architectural surveying and took measurements on site. In the second week, after returning to Beijing, the students drew up architectural drawings on the computer. On-site measurement tools included total station, laser rangefinder, leather and steel measuring tape, level ruler, triangle ruler, and for some parts, a temporary structure and scaffolding. We produced a total of two hundred and seventy-six drawings of historical buildings (one hundred and forty-nine of Zhongyue Temple, eighty-five of Shaolin Monastery, and forty-two of Songyang Academy). This volume of the *Traditional Chinese Architecture Surveying and Mapping Series* presents two hundred and thirty-nine drawings of the general site layout, elevation and section as well as on floor plan, elevation, and section of individual buildings and components. The drawings were not only crucial for the application of nomination of the historic monuments of Songshan as World Cultural Heritage Sites, but they provide the institutions and bureaus engaged in cultural heritage protection with important information, establishing a pool of reliable and relevant first-hand data for future preservation and restoration efforts of historical buildings (Fig.33).

Tsinghua professors (WANG Guixiang, LIAO Huinong, LIU Chang, and HE Congrong) and graduate students (LI Luke, JIANG Dongcheng, YANG Bo, AO Shiheng, XIN Huiyuan, DUAN Zhijun, XU Xiaoying, MEI Jing, LI Dehua, and ZHENG Liang) led several small teams of undergraduate students (in total eighty-four students; Class of 2004) that were responsible for the basic surveying and mapping work on site.

严整的建筑布局使中岳庙成为一组主次分明、错落有致、布局紧凑、色调和谐的庞大建筑群，是五岳庙中现存规模最大、保存较完整的一组，也是河南省规模最巨、最完整的古建筑群，是中国古代礼制建筑格局最全面的代表。

三、测绘和编辑工作简介

本次测绘历经两周，第一周在现场测绘，第二周回到学校进行计算机绘图。测量工具主要包括全站仪、激光测距仪、皮尺、水平尺、三角尺、钢卷尺等，局部搭建脚手架。共完成古建筑测绘图纸 276 张（中岳庙 149 张、少林寺 85 张、嵩阳书院 42 张），本册测绘图集辑选了其中的 239 张。图纸内容主要包括建筑群的总平面、总立面、总剖面图，重要建筑单体的平面、立面、剖面和细部大样图。这些测绘图不仅为古建筑管理部门提供了精细的实测图纸，为文物建筑的保护与修缮留下了一份珍贵的第一手资料，同时也为河南嵩山古建筑群申请世界文化遗产提供了重要的基础性资料支持（图 33）。

这次测绘，由王贵祥、廖慧农、刘畅、贺从容等老师带队辅导，参加辅导的研究生有李路珂、姜东成、杨博、敖仕恒、辛惠园、段智君、徐晓颖、梅静、李德华、郑亮等。承担主要测绘任务的是清华大学建筑学院 2004 级 84 位本科生。

图 33　测绘工作照　图片来源：廖慧农 提供

Fig.33 Work photos Source: Provided by LIAO Huinong

The intensive two-week program offered students an opportunity to become immersed in the mapping of historical architecture and gain first-hand practical experience. In spite of the heat, they climbed up the mountain, not afraid to get their hands dirty, they explored dusty roof structures, and sparing no efforts, they used all appropriate means to perform the tasks at hand. The names of the instructors and students responsible for measuring and mapping are indicated on the drawings. The fieldwork would not have been possible without the help of the local administration. I would like to thank REN Wei, WANG Wenhua, and ZHANG Jianhua from the Zhengzhou Municipal Cultural Bureau and Jin Yindong and Gong Songtao from the Dengfeng Cultural Relics Bureau.

Finally, Tang Henglu from Wang Guixiang Studio was responsible for the editing of this book and organized, modified and supplemented the students' drawings together with his colleagues MAI Linlin, SHAN Menglin, and HU Jingfu. Tsinghua professor LIAO Huinong and Tsinghua graduate YANG Bo who also participated in the fieldwork provided valuable pictures for this volume of the *Traditional Chinese Architecture Surveying and Mapping Series*. Tsinghua graduate student HE Wenxuan was responsible for the illustrations of the preface. This book is the result of their joint effort and I owe them a great deal of gratitude.

这两周他们全身心投入到古建筑测绘的指导与实习中，为完整地描绘在烈日下爬高，为获取准确数据在积灰的梁架间穿行，为测绘质量付出了辛勤劳动。每份测绘图上，也留下了绘图者、测量者和辅导教师的名字。本次测绘工作还得到了郑州市文化局任伟、于文华、张建华等同志，以及登封市文物局靳银东、宫嵩涛等同志的大力支持和帮助，没有他们热情尽责的支持与帮助，我们测绘中遇到的种种困难无法得到很好的解决和克服，在此表示由衷感谢。

最后，关于本册图集的整理修改工作，主要是由王贵祥教授工作室的唐恒鲁负责，他和清华同衡的买琳琳、单梦林及胡竞芙，在当年学生测绘图的基础上，对每份图纸进行修改、整理、补充与初步排版，为图面的完善付出了辛苦。廖慧农老师和杨博博士，既是本次测绘的参与者，也为图册提供了宝贵的图片。研究生何文轩，认真承担了前言的插图工作。这本图集得以呈现，离不开众多同仁的付出和支持，在此一并表示感谢。

图

版

Figure

少林寺
Shaoling Monastery

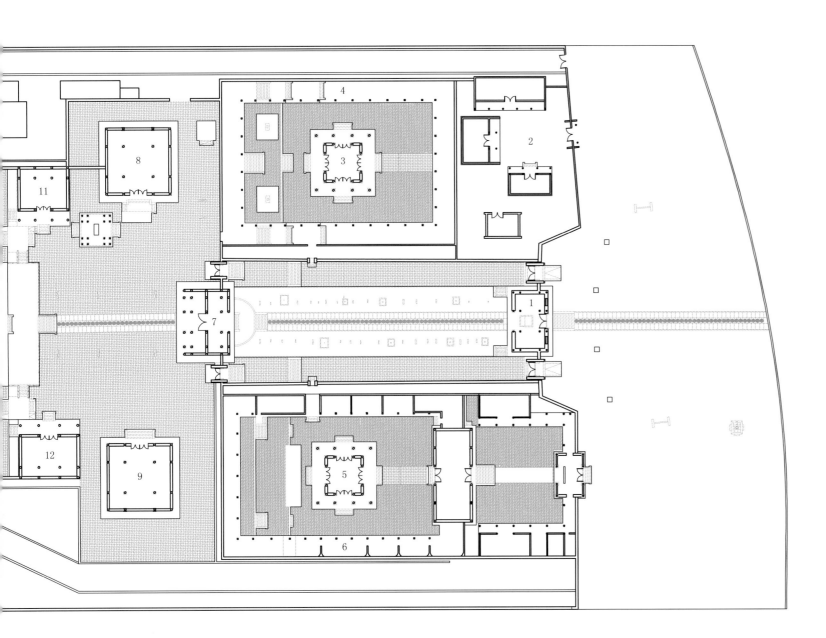

1　山门
2　民宅
3　慈云堂
4　碑廊
5　西来堂
6　展廊
7　天王殿
8　钟楼
9　鼓楼
10　大雄宝殿
11　那罗殿
12　六祖殿
13　藏经阁
14　禅堂
15　方丈室
16　客堂
17　立雪亭
18　文殊殿
19　普贤殿
20　千佛殿
21　观音殿
22　地藏殿
23　练功房
24　僧舍
25　宋塔

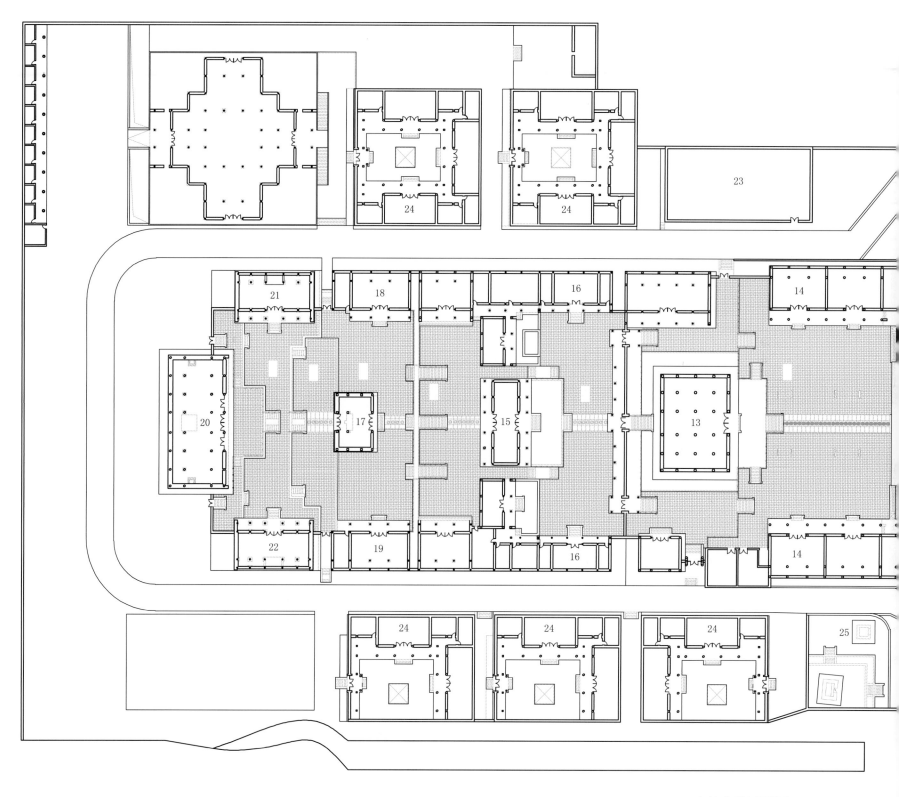

少林寺总平面图
Site plan of Shaolin Monastery

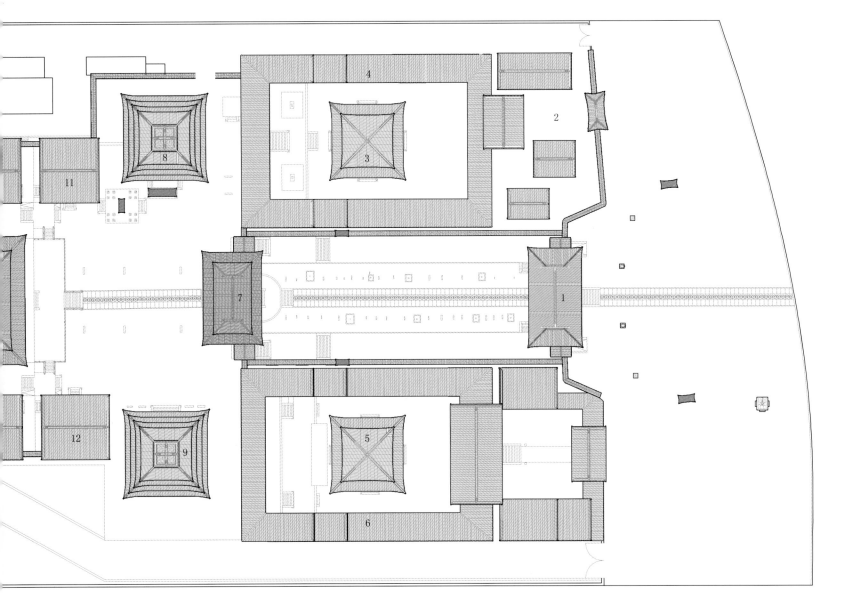

1 山门
2 民宅
3 慈云堂
4 碑廊
5 西来堂
6 展廊
7 天王殿
8 钟楼
9 鼓楼
10 大雄宝殿
11 那罗殿
12 六祖殿
13 藏经阁
14 禅堂
15 方丈室
16 客堂
17 立雪亭
18 文殊殿
19 普贤殿
20 千佛殿
21 观音殿
22 地藏殿
23 练功房
24 僧舍
25 宋塔

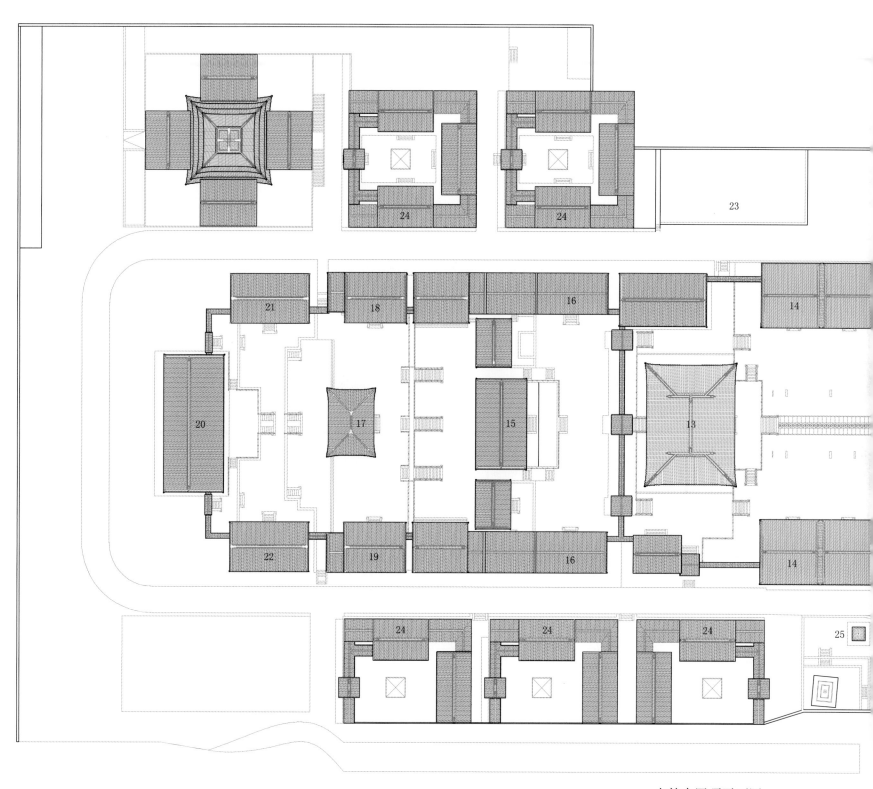

少林寺屋顶平面图
Roof plan of Shaolin Monastery

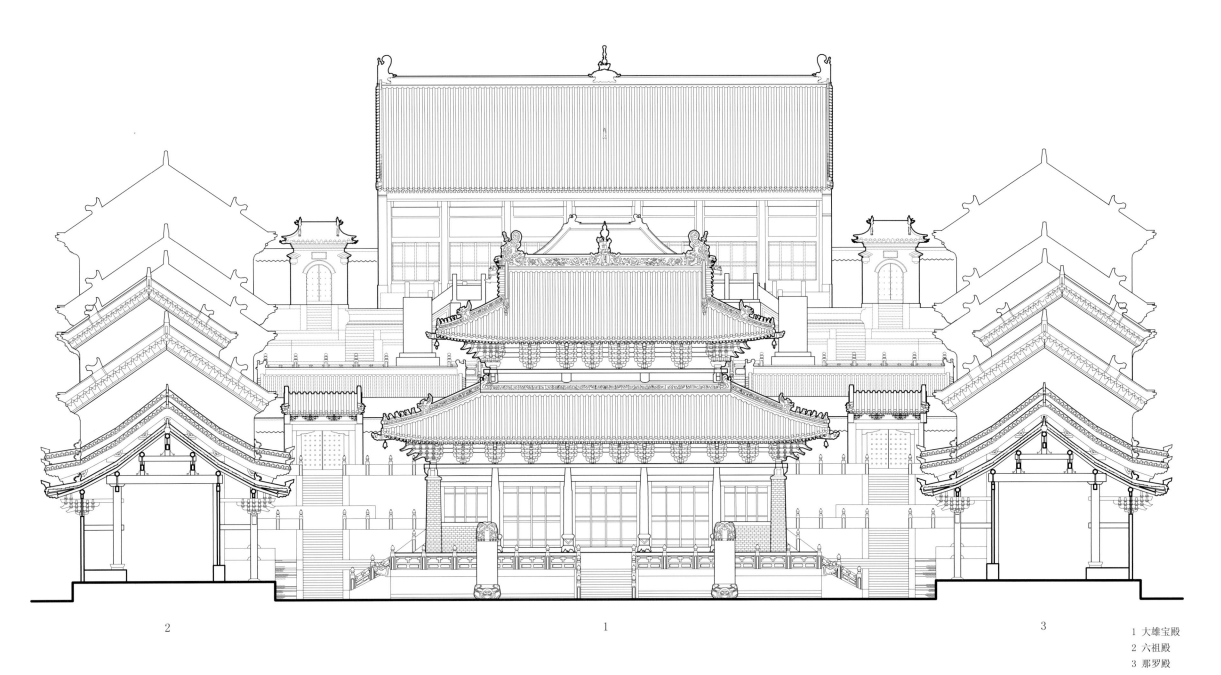

2

1

3

1　大雄宝殿
2　六祖殿
3　那罗殿

0　1　　　5m

少林寺大雄宝殿前总立面图
Site elevation of front of Daxiong baodian, Shaolin Monastery

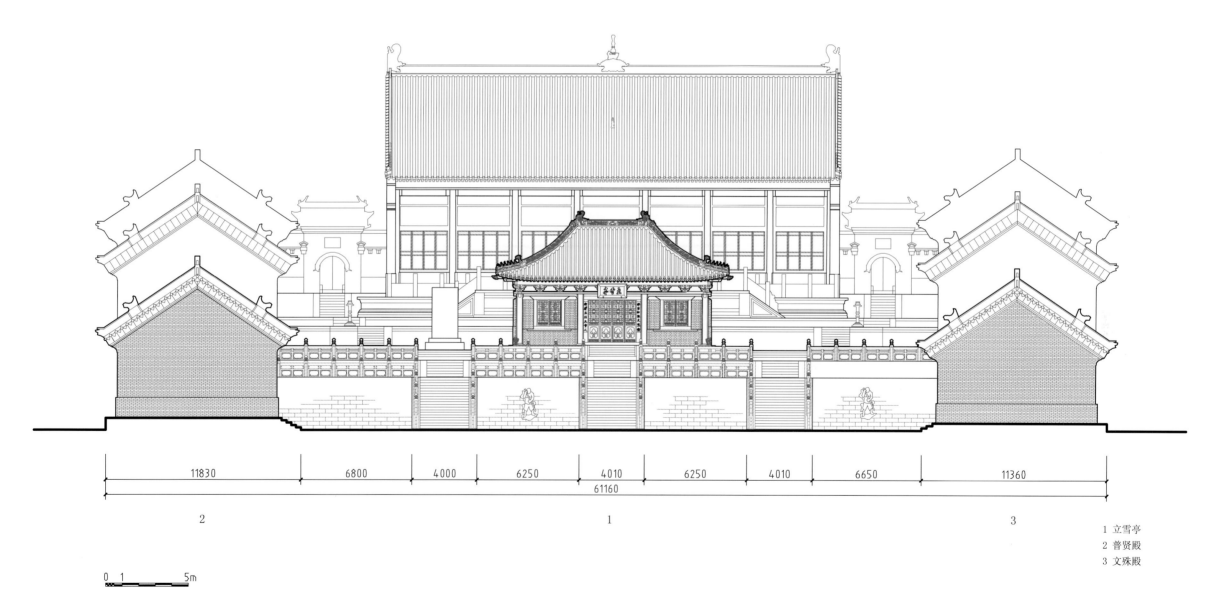

| 11830 | 6800 | 4000 | 6250 | 4010 | 6250 | 4010 | 6650 | 11360 |

61160

2　　　　　　　　　　　　　　　1　　　　　　　　　　　　　　　3

1　立雪亭
2　普贤殿
3　文殊殿

0　1　　　　5m

少林寺立雪亭前总立面图
Site elevation of front of Lixueting, Shaolin Monastery

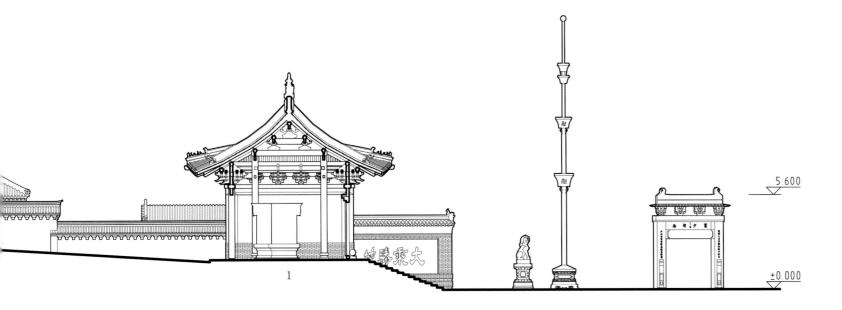

1

5 600

±0 000

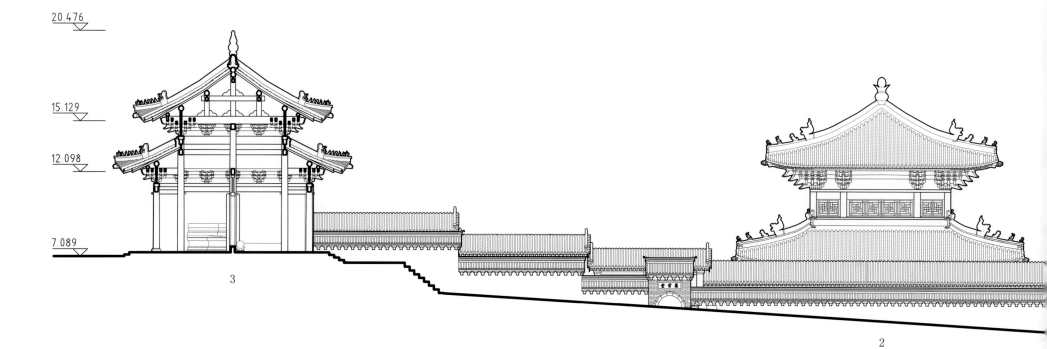

20.476

15.129

12.098

7.089

3

2

0 1   5m

少林寺山门前广场至天王殿纵剖面图

Longitudinal section from front square of *shanmen* to Tianwangdian, Shaolin Monastery

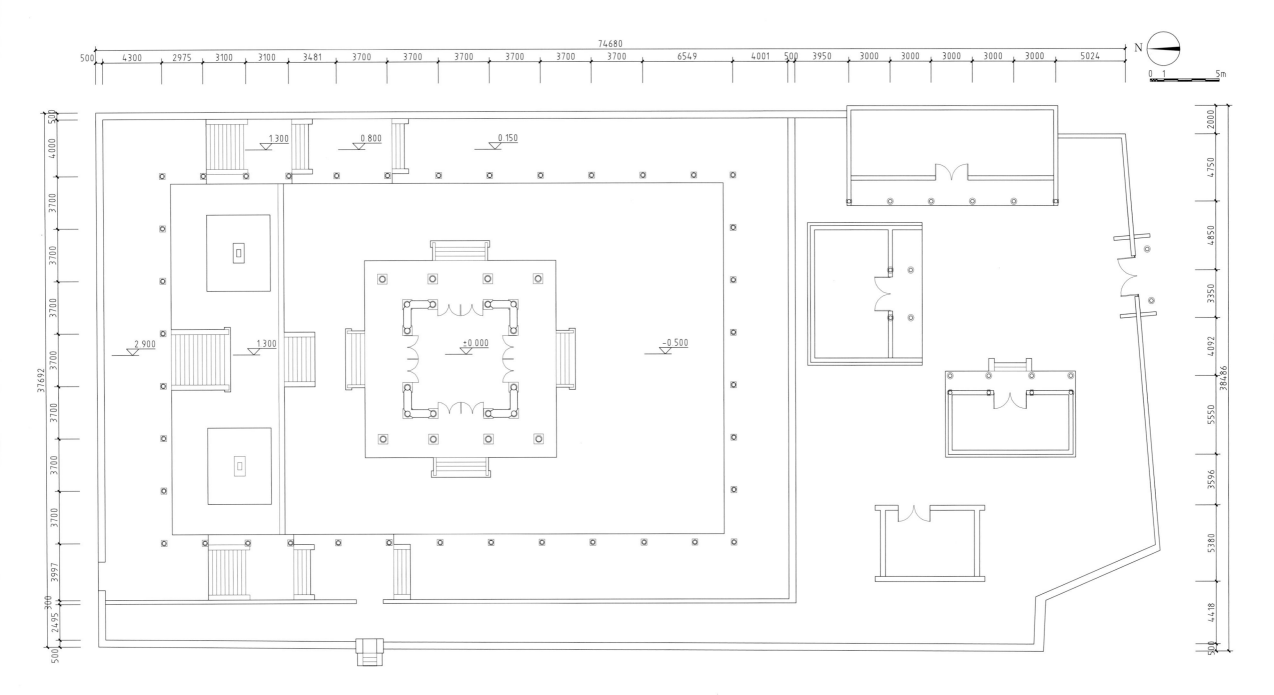

少林寺天王殿西院平面图
Plan of Xiyuan of Tianwangdian, Shaolin Monastery

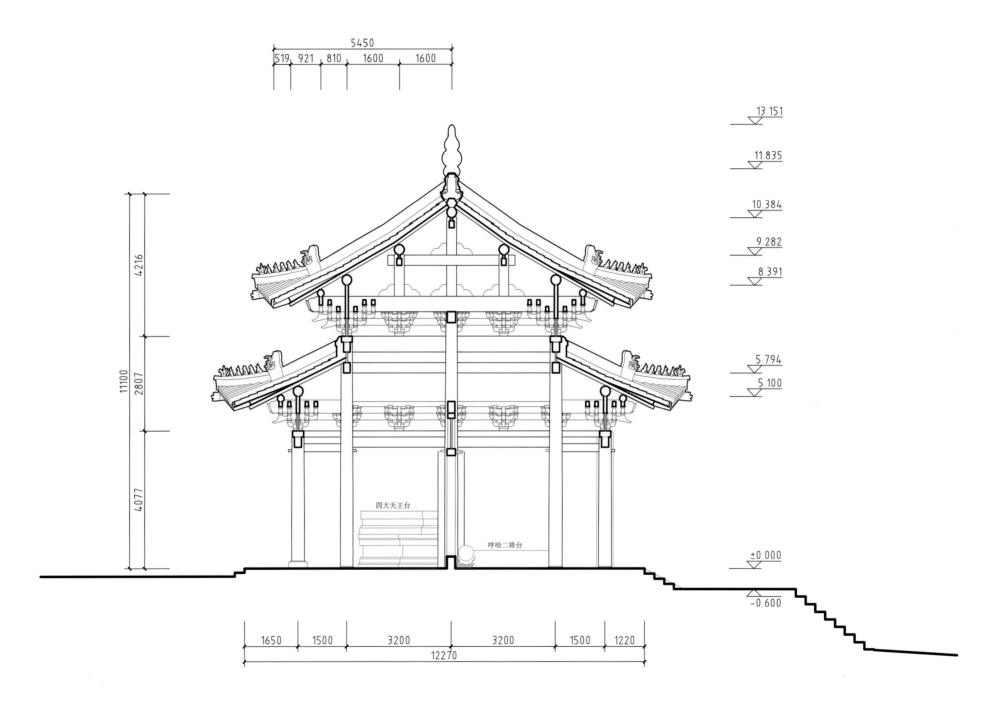

5450
519 921 810 1600 1600

13.151
11.835
10.384
9.282
8.391

5.794
5.100

±0.000
−0.600

4216
11100
2807
4077

四大天王台
哼哈二将台

1650 1500 3200 3200 1500 1220
12270

0 1 2m

少林寺天王殿横剖面图
Cross-section of Tianwangdian of Shaolin Monastery

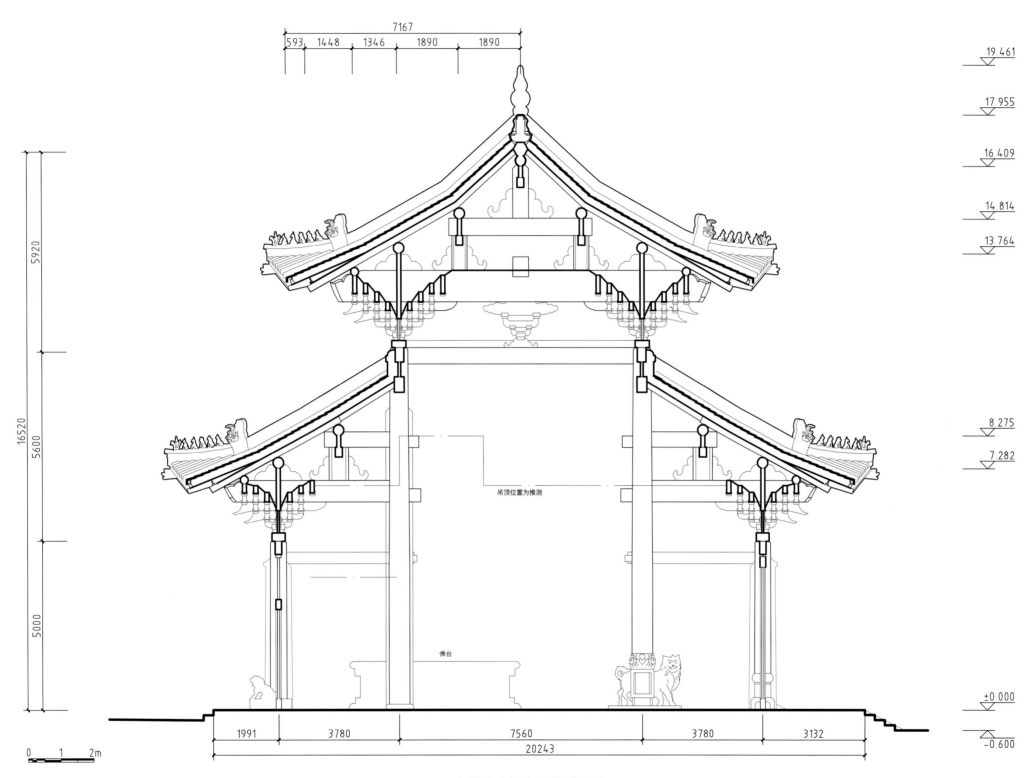

吊顶位置为推测

佛台

少林寺大雄宝殿横剖面图

Cross-section of Daxiong baodian of Shaolin Monastery

0 1 2m

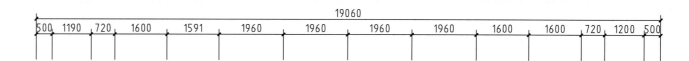
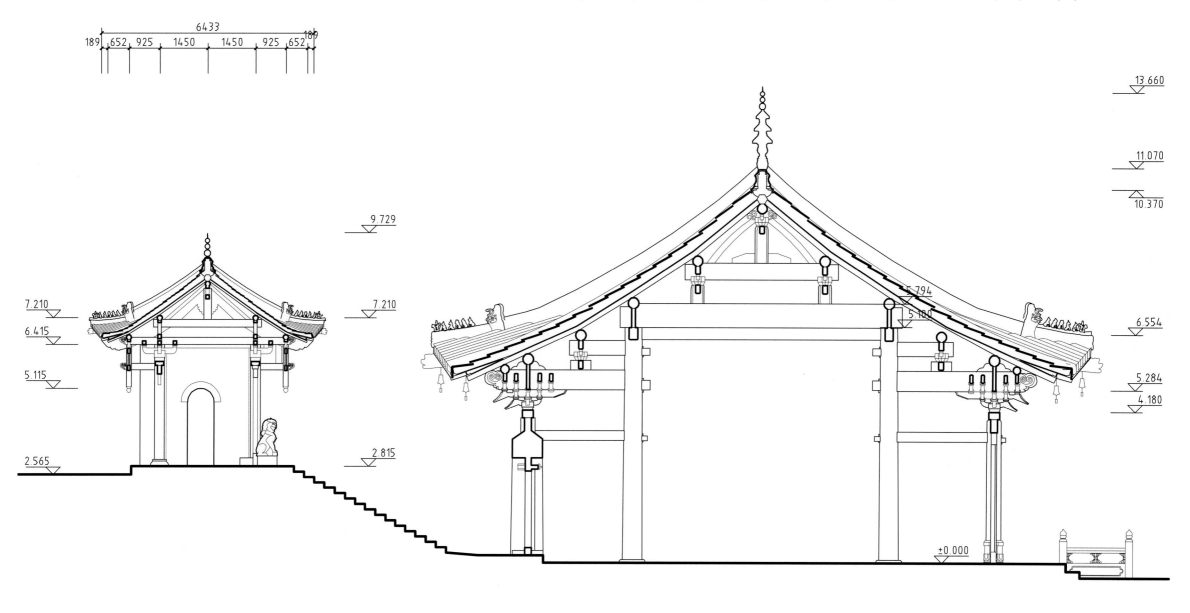

0 1 2m

少林寺藏经阁横剖面图
Cross-section of Cangjingge of Shaolin Monastery

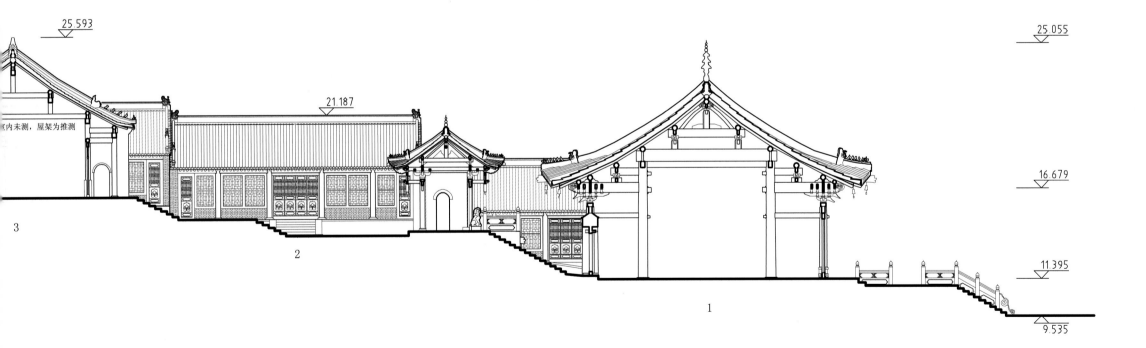

25.593

25.055

21.187

室内未测，屋架为推测

16.679

3

2

11.395

1

9.535

1　藏经阁
2　客堂
3　方丈室
4　立雪亭
5　观音殿
6　千佛殿

37.572

30.935

32.368

27.638

27.523

22.608

23.482

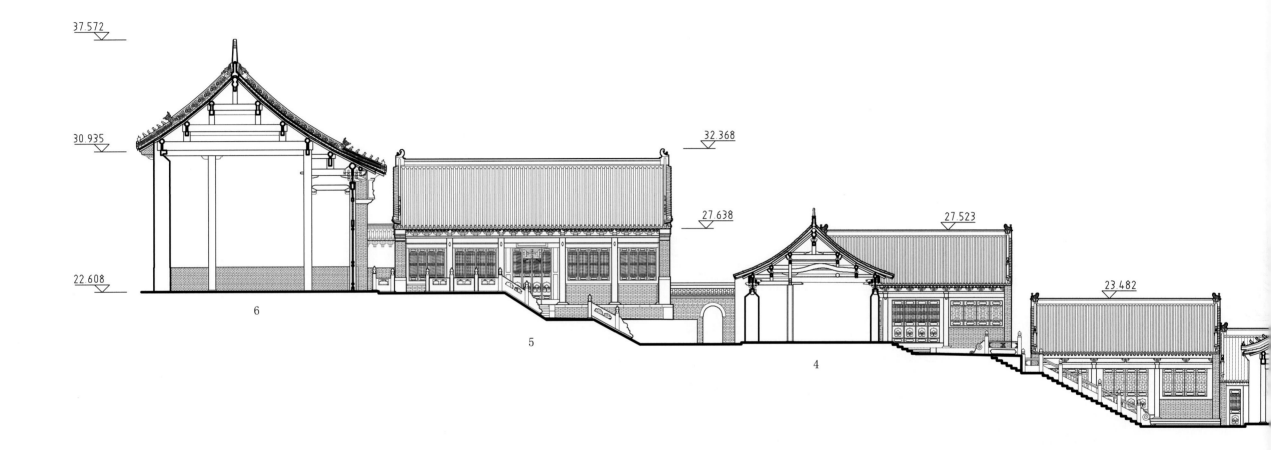

6

5

4

0 1   5m

少林寺藏经阁至千佛殿纵剖面图
Longitudinal section from Cangjingge to Qianfodian, Shaolin Monastery

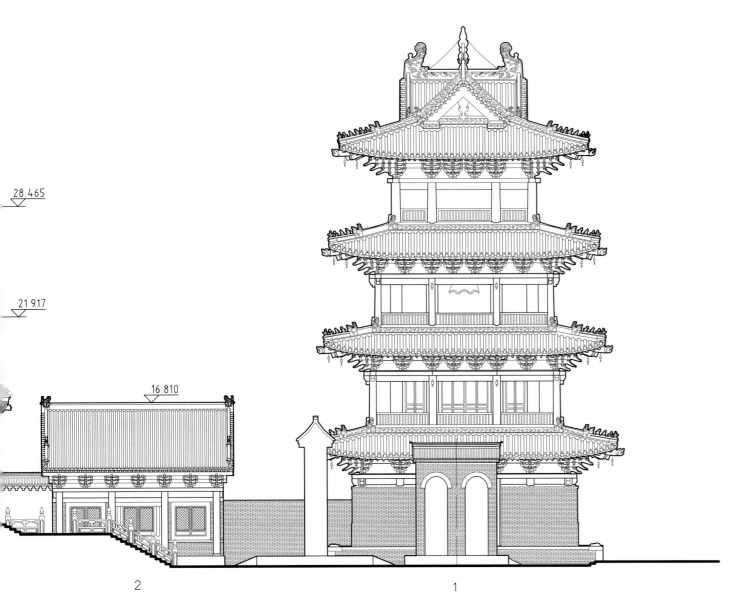

28.465

21.917

16.810

2

1

1 钟楼
2 那罗殿
3 大雄宝殿
4 禅堂

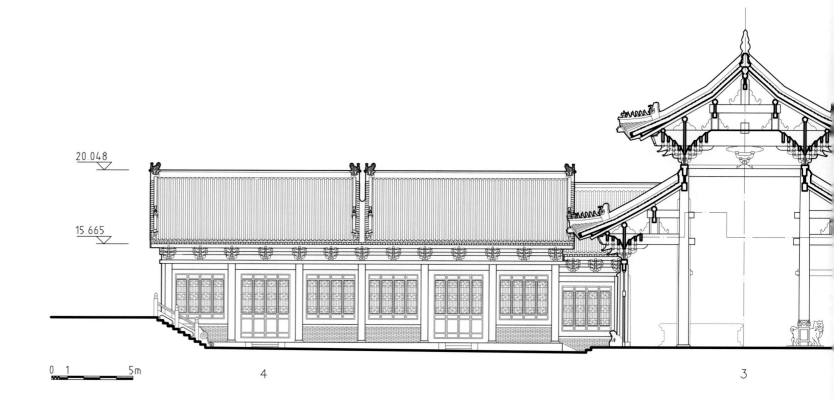

20.048

15.665

0 1 5m

4

3

少林寺天王殿至藏经阁纵剖面图
Longitudinal section from Tianwangdian to Cangjingge, Shaolin Monastery

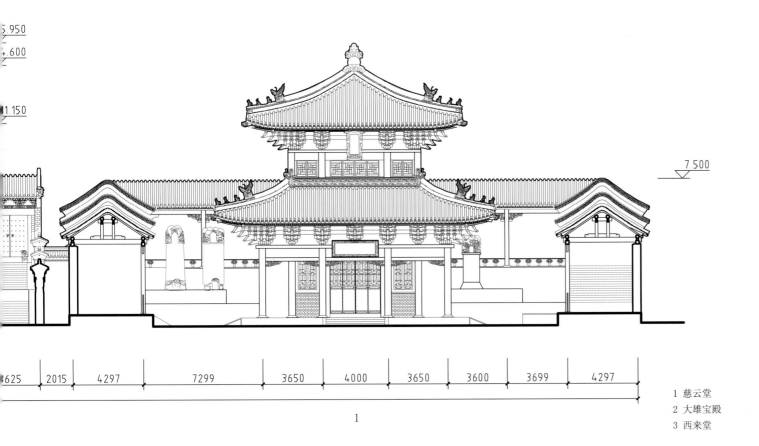

5.950
+600
1.150
7.500

625 | 2015 | 4297 | 7299 | 3650 | 4000 | 3650 | 3600 | 3699 | 4297

1

1　慈云堂
2　大雄宝殿
3　西来堂

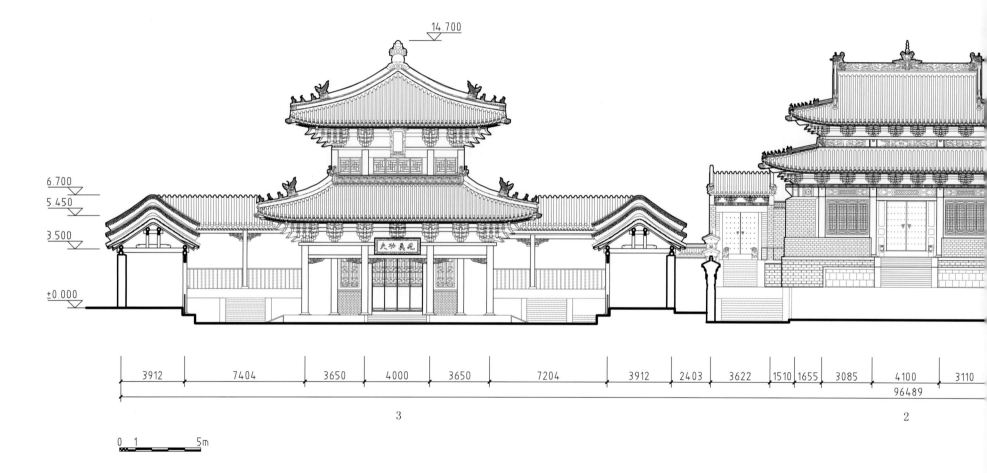

14.700

6.700
5.450
3.500

±0.000

3912　7404　3650　4000　3650　7204　3912　2403　3622　1510 1655　3085　4100　3110

96489

3

2

0　1　　　5m

少林寺天王殿前中轴横剖面图
Cross-section of front axis of Tianwangdian, Shaolin Monastery

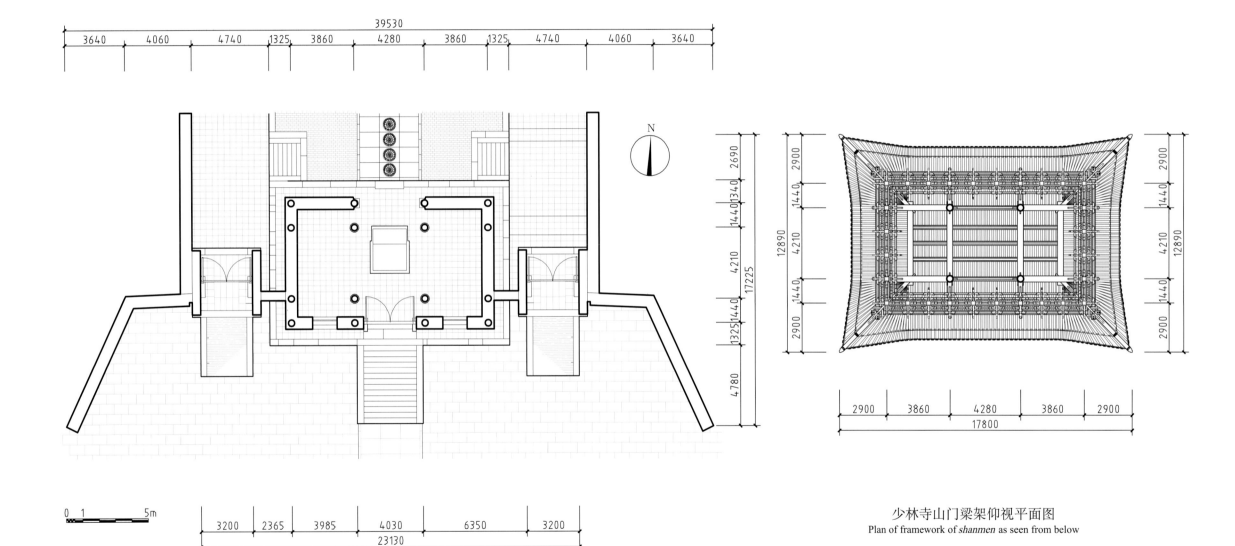

少林寺山门平面图
Plan of *shanmen* of Shaolin Monastery

少林寺山门梁架仰视平面图
Plan of framework of *shanmen* as seen from below

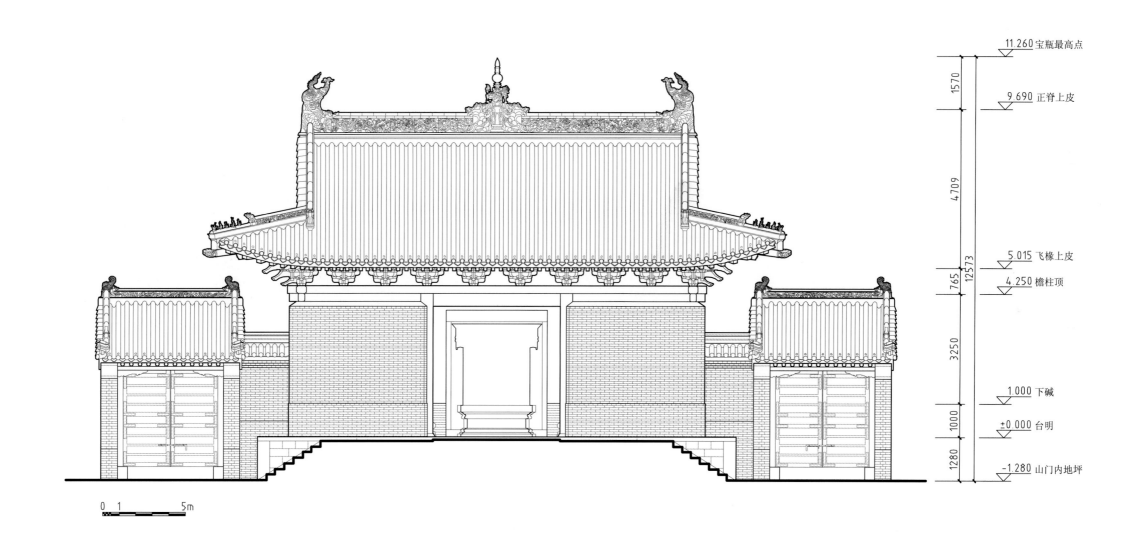

11.260 宝瓶最高点

9.690 正脊上皮

5.015 飞椽上皮

4.250 檐柱顶

1.000 下碱

±0.000 台明

-1.280 山门内地坪

1570

4709

12573

765

3250

1000

1280

少林寺山门背立面图
Rear elevation of *shanmen* of Shaolin Monastery

0 1 5m

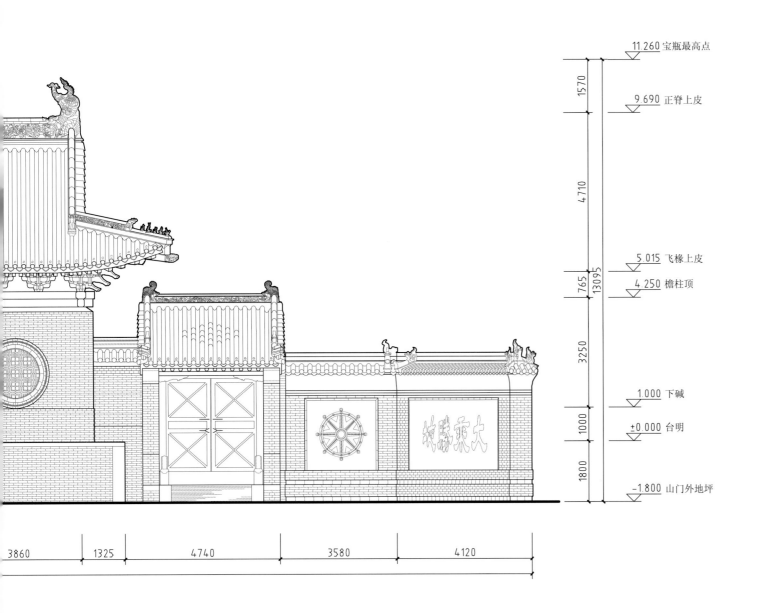

11.260 宝瓶最高点

9.690 正脊上皮

5.015 飞椽上皮

4.250 檐柱顶

1.000 下碱

±0.000 台明

-1.800 山门外地坪

1570

4710

765

3095

13095

3250

1000

1800

大乘胜地

3860   1325   4740   3580   4120

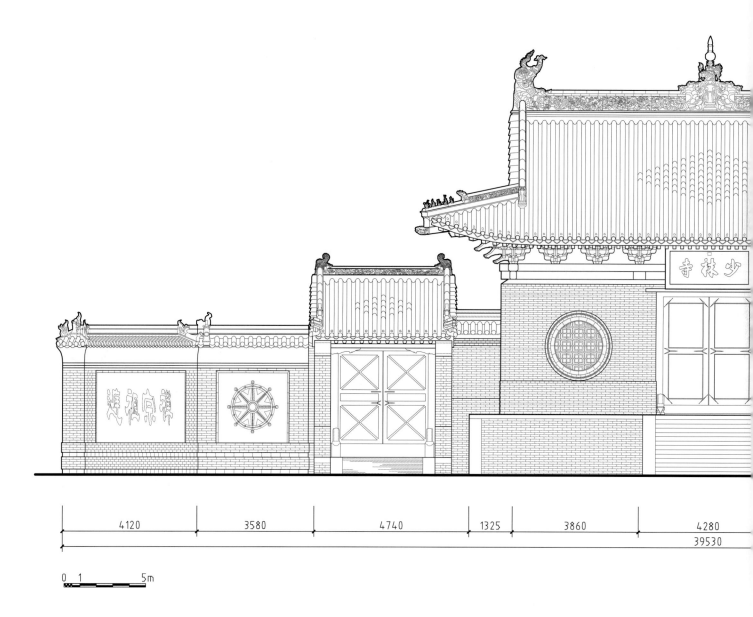

4120	3580	4740	1325	3860	4280

39530

0 1 5m

少林寺山门正立面图
Front elevation of *shanmen* of Shaolin Monastery

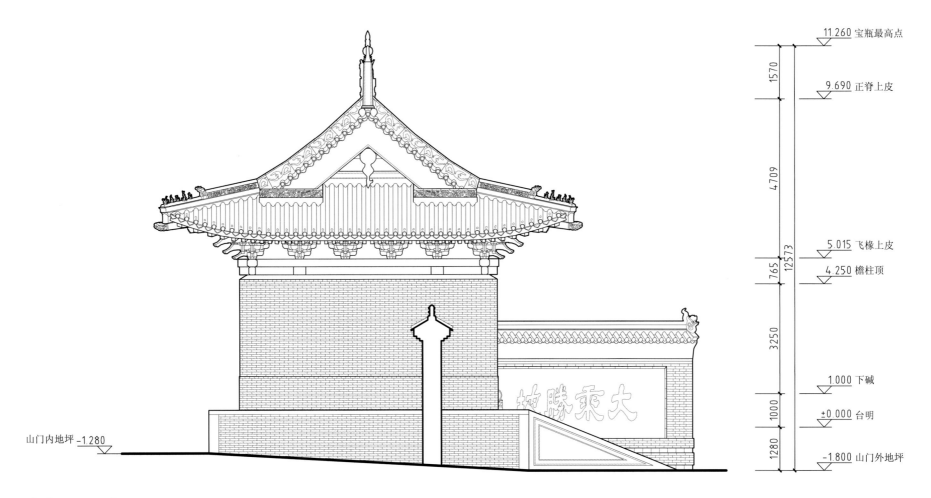

11.260 宝瓶最高点

9.690 正脊上皮

1570

4709

12573

5.015 飞椽上皮

765

4.250 檐柱顶

3250

1.000 下碱

1000

±0.000 台明

1280

-1.800 山门外地坪

山门内地坪 -1.280

0 1 5m

少林寺山门侧立面图
Side elevation of *shanmen* of Shaolin Monastery

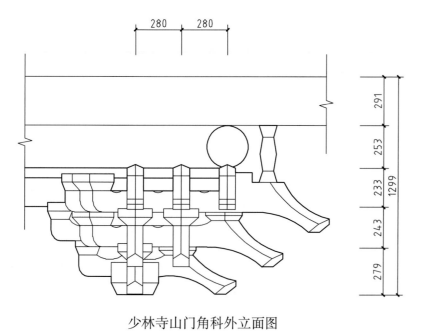

少林寺山门角科外立面图
Elevation of outer *jiaoke* of *shanmen*, Shaolin Monastery

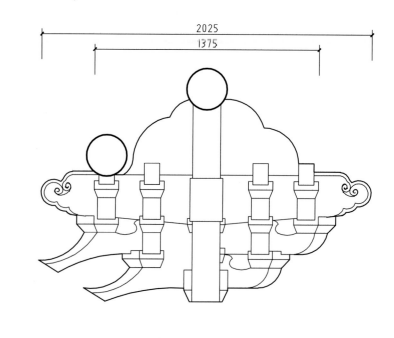

少林寺山门柱头科正立面图
Front elevation of *zhutouke* of *shanmen*, Shaolin Monastery

少林寺山门柱头科侧立面图
Side elevation of *zhutouke* of *shanmen*, Shaolin Monastery

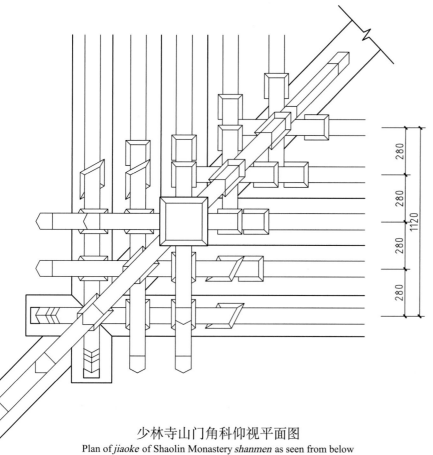

少林寺山门角科仰视平面图
Plan of *jiaoke* of Shaolin Monastery *shanmen* as seen from below

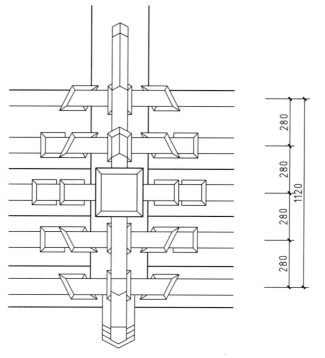

少林寺山门柱头科仰视平面图
Plan of *zhutouke* of Shaolin Monastery *shanmen* as seen from below

0   0.5   1m

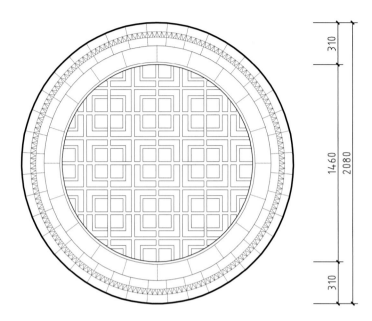

少林寺山门立面花窗大样图
Elevation of *huachuang* of *shanmen*

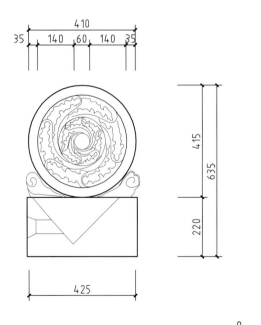

少林寺山门屋脊剖面大样图
Section of *wuji* of *shanmen*, Shaolin Monastery

少林寺山门抱鼓石大样图
Elevation of *baogushi* of *shanmen*, Shaolin Monastery

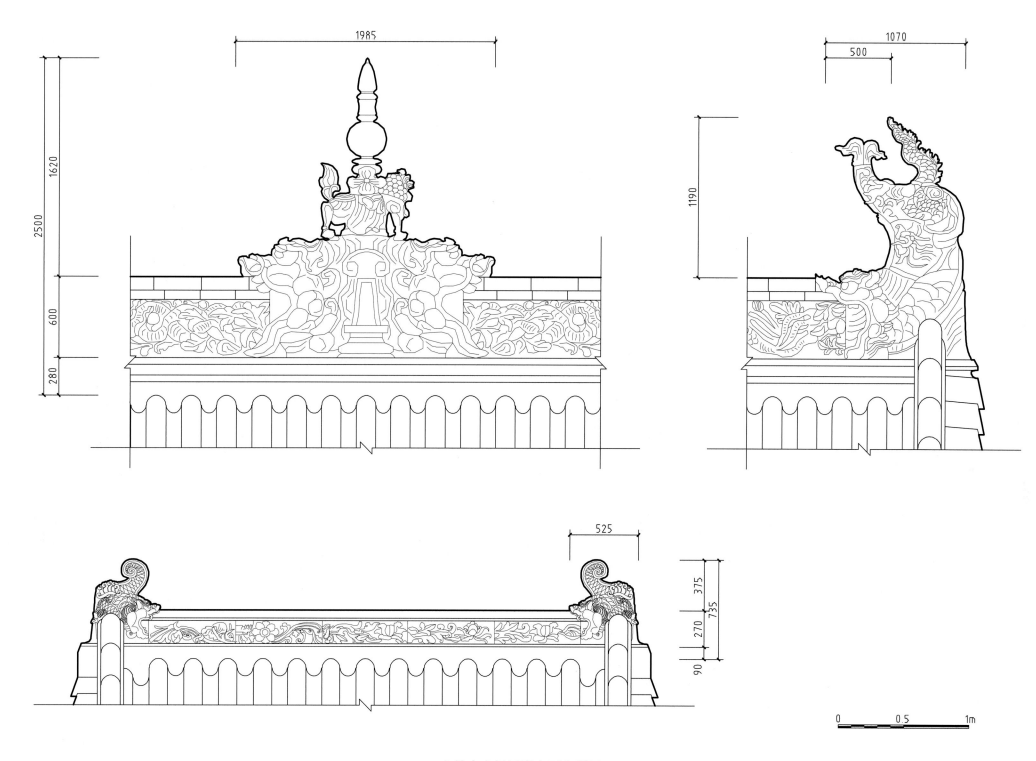

少林寺山门屋脊立面大样图
Elevation of *wuji* of *shanmen*, Shaolin Monastery

0　0.5　1m

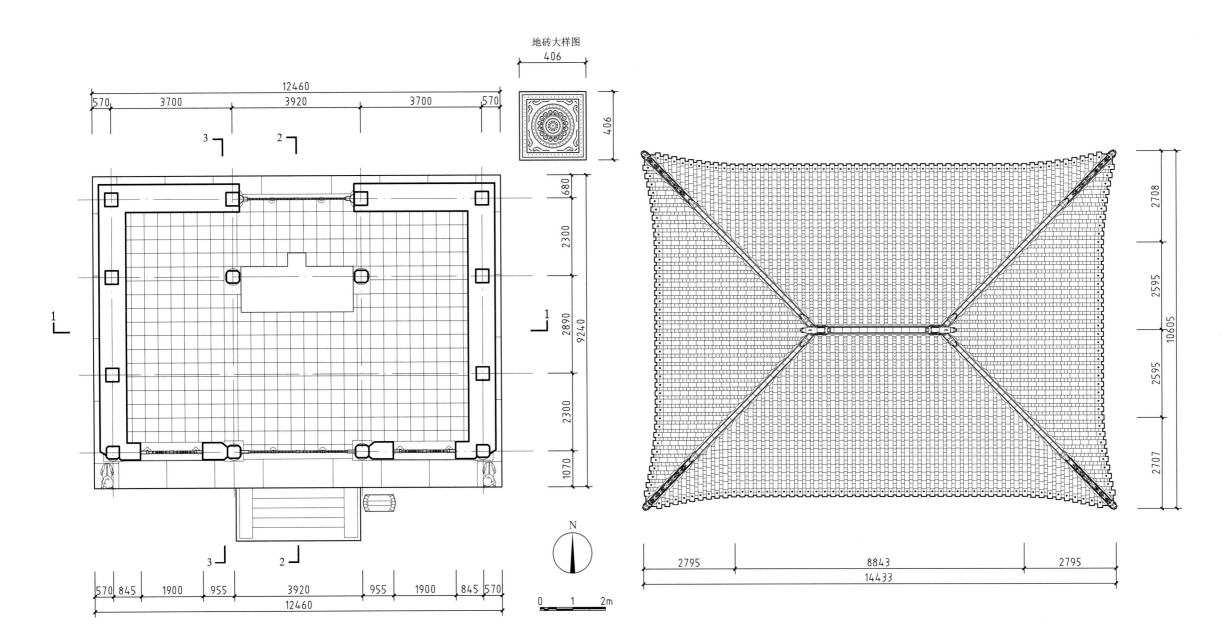

地砖大样图

少林寺立雪亭首层平面图
Plan of first floor of Lixueting, Shaolin Monastery

少林寺立雪亭屋顶平面图
Roof plan of Lixueting, Shaolin Monastery

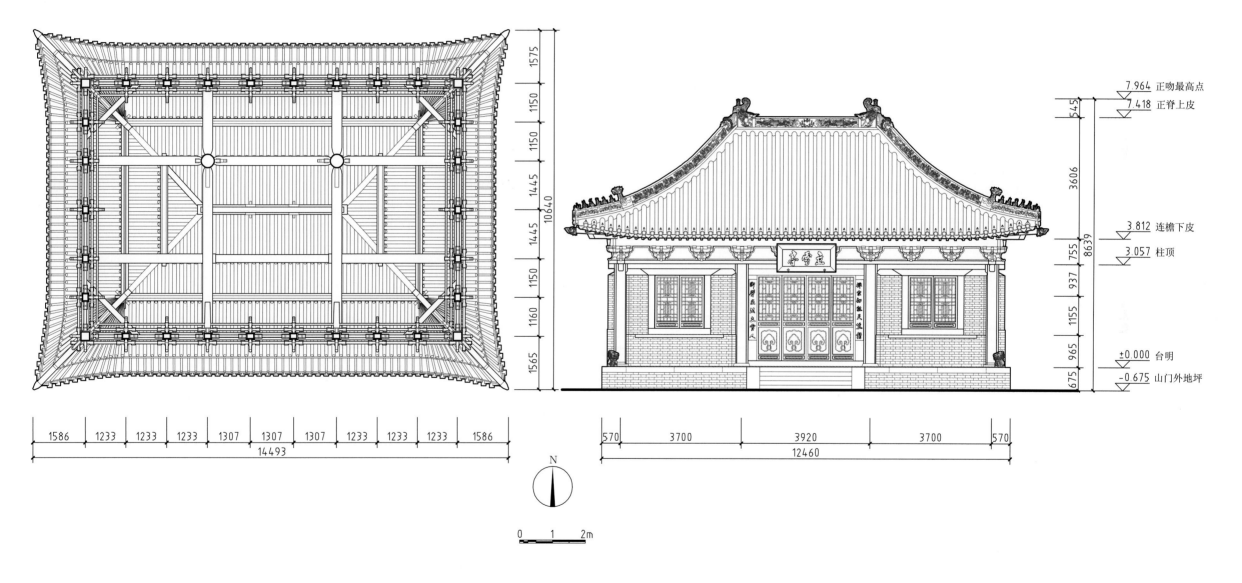

少林寺立雪亭梁架仰视平面图
Plan of framework of Shaolin Monastery Lixueting as seen from below

少林寺立雪亭正立面图
Front elevation of Lixueting of Shaolin Monastery

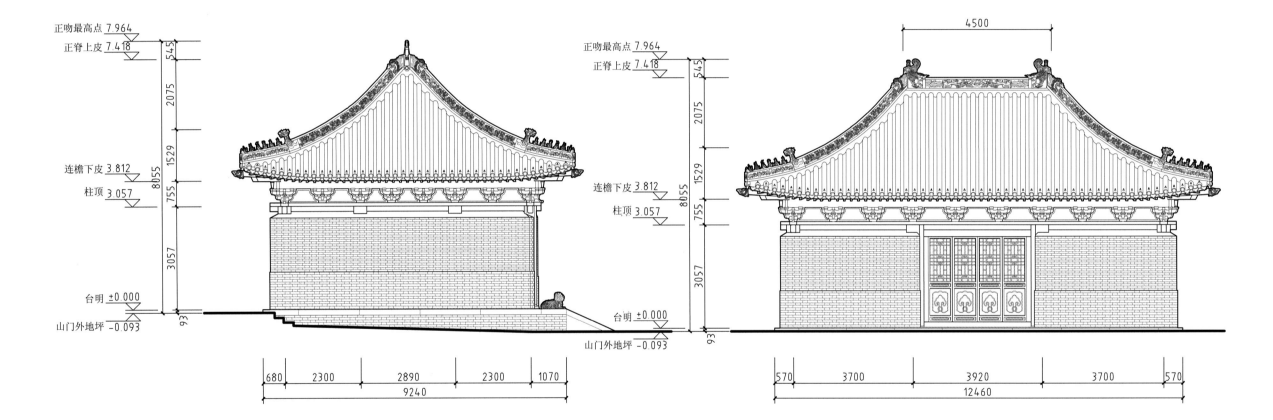

正吻最高点 7.964
正脊上皮 7.418
连檐下皮 3.812
柱顶 3.057
台明 ±0.000
山门外地坪 -0.093

545
2075
1529
755
3057
8055
93

680 2300 2890 2300 1070
9240

正吻最高点 7.964
正脊上皮 7.418
连檐下皮 3.812
柱顶 3.057
台明 ±0.000
山门外地坪 -0.093

4500

545
2075
1529
755
3057
8055
93

570 3700 3920 3700 570
12460

0 1 2m

少林寺立雪亭侧立面图
Side elevation of Lixueting of Shaolin Monastery

少林寺立雪亭背立面图
Rear elevation of Lixueting of Shaolin Monastery

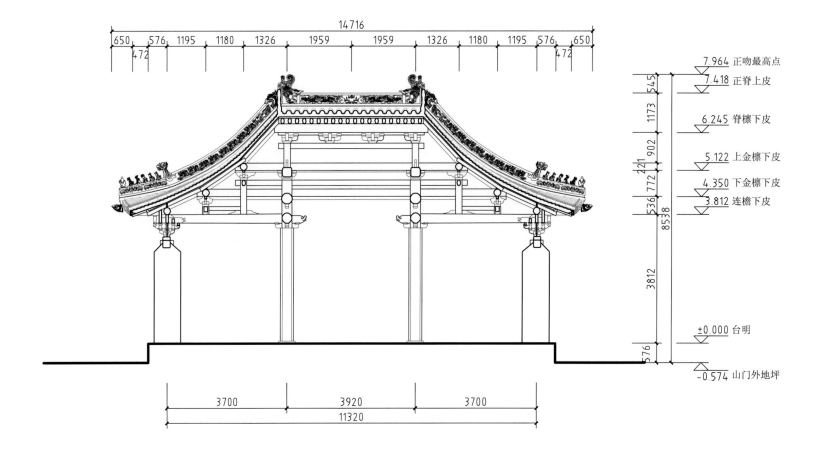

14716

| 650 | 576 | 1195 | 1180 | 1326 | 1959 | 1959 | 1326 | 1180 | 1195 | 576 | 650 |

472

472

7.964 正吻最高点
7.418 正脊上皮
545
1173
6.245 脊檩下皮
902
221
5.122 上金檩下皮
772
4.350 下金檩下皮
536
3.812 连檐下皮
8538
3812
±0.000 台明
576
-0.574 山门外地坪

| 3700 | 3920 | 3700 |
11320

少林寺立雪亭 1-1 剖面图
Section 1-1 of Lixueting of Shaolin Monastery

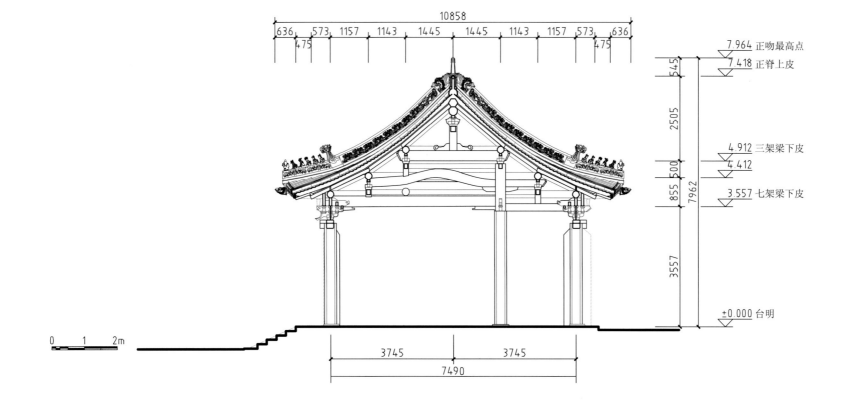

少林寺立雪亭 2-2 剖面图
Section 2-2 of Lixueting of Shaolin Monastery

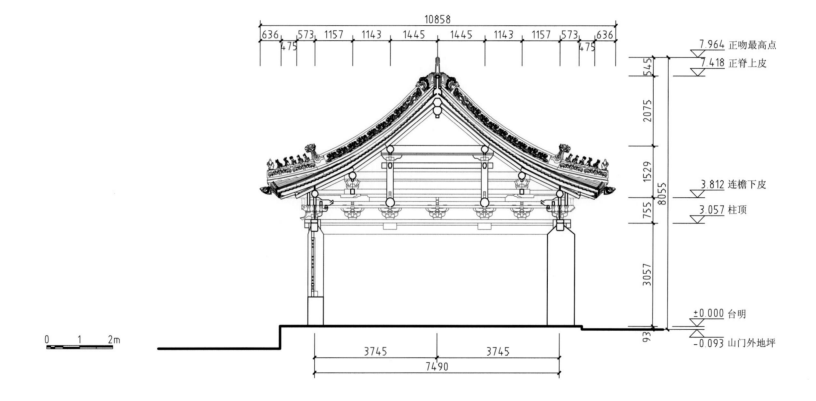

10858

636 573 1157 1143 1445 1445 1143 1157 573 636

475 475

7.964 正吻最高点
7.418 正脊上皮

545

2075

1529

8055

3.812 连檐下皮

755

3.057 柱顶

3057

±0.000 台明

93

-0.093 山门外地坪

0 1 2m

3745 3745

7490

少林寺立雪亭 3-3 剖面图
Section 3-3 of Lixueting of Shaolin Monastery

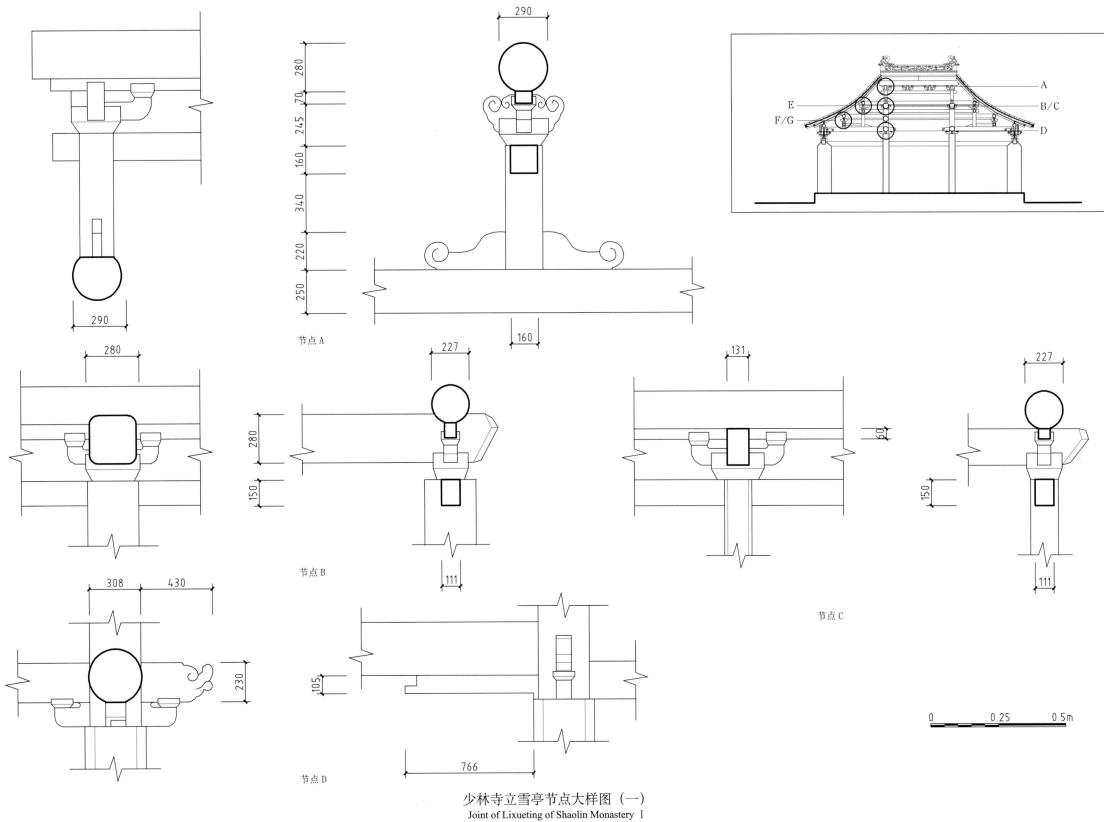

290

280

70

245

160

340

220

250

节点 A

160

290

280

节点 B

280

227

150

111

131

60

节点 C

227

150

111

308

430

230

105

766

节点 D

E

F/G

A

B/C

D

0   0.25   0.5m

少林寺立雪亭节点大样图（一）
Joint of Lixueting of Shaolin Monastery Ⅰ

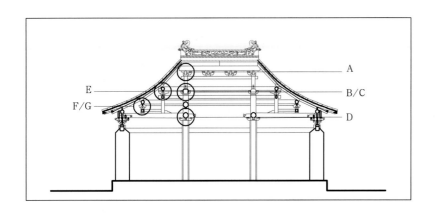

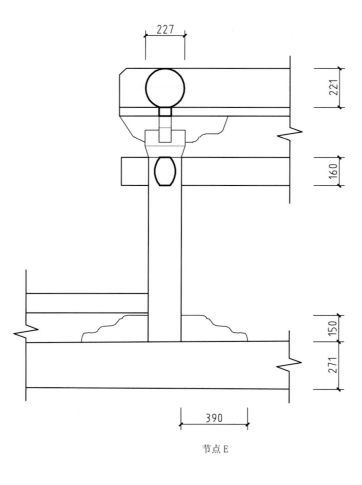

节点 E

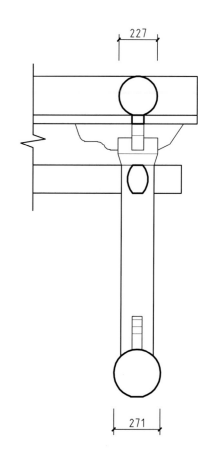

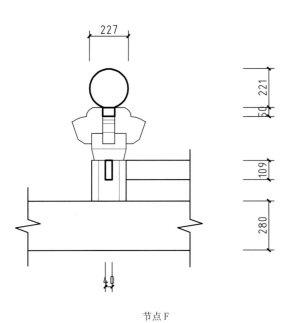

节点 F

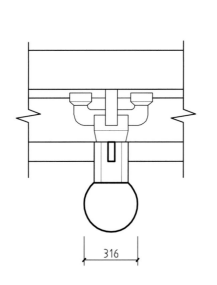

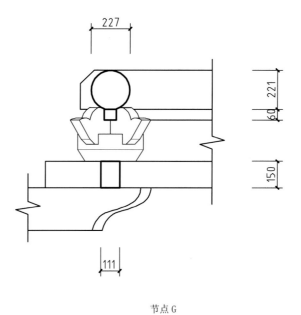

节点 G

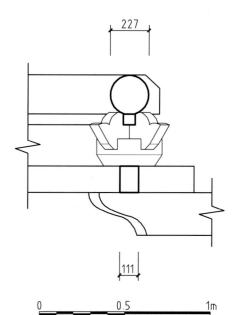

0　　　　0.5　　　　1m

少林寺立雪亭节点大样图（二）
Joint of Lixueting of Shaolin Monastery Ⅱ

I 型

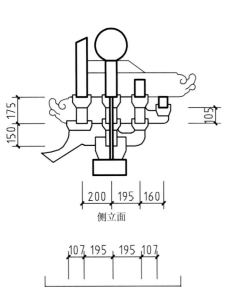

150 175

200 195 160

侧立面

107 195 195 107

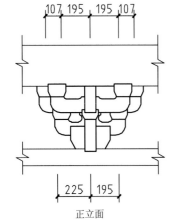

225 195

正立面

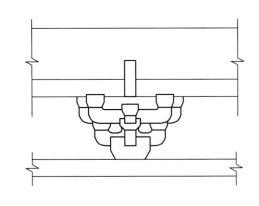

背立面

II 型

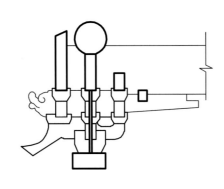

105

侧立面

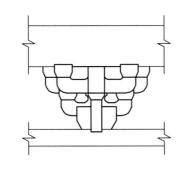

正立面

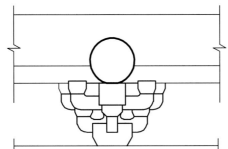

背立面

IV 型（含 IV'' 型）

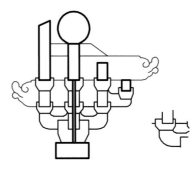

侧立面

注：IV'' 型标示散斗为五边形平面

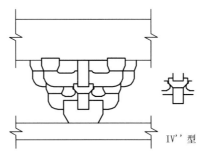

IV'' 型

正立面

注：IV 型和 IV'' 型背立面同 I 型

IV'' 型

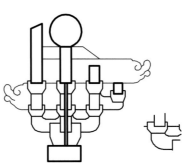

少林寺立雪亭平身科、柱头科斗栱大样图（一）
Bracket set of *pingshenke* and *zhutouke* of Lixueting, Shaolin Monastery I

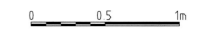

0　　0.5　　1m

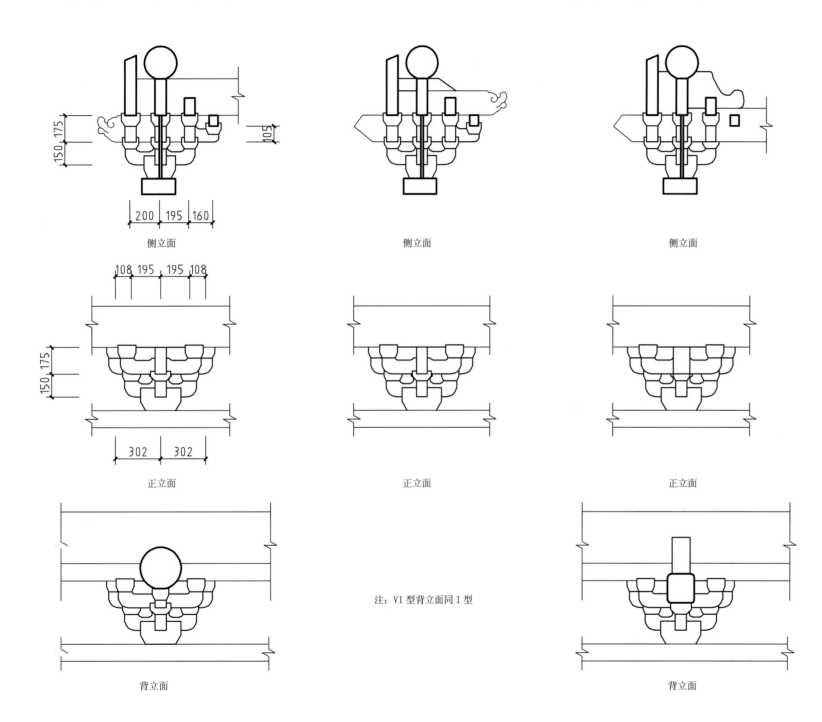

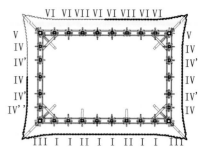

IV′型　　　　　　　VI型　　　　　　　VII型

105

150 175

200 195 160

侧立面　　　　　　侧立面　　　　　　侧立面

108 195 195 108

150 175

302 302

正立面　　　　　　正立面　　　　　　正立面

注：VI型背立面同I型

背立面　　　　　　　　　　　　背立面

065

0　　　0.5　　　1m

少林寺立雪亭平身科、柱头科斗栱大样图（二）
Bracket set of *pingshenke* and *zhutouke* of Lixueting, Shaolin Monastery II

III 型

V 型

150 175

200 195

150 175

侧立面

侧立面

VI VI VII VI VI VII VI VI

V V
IV IV
IV' IV'
IV IV
IV' IV'
IV'' IV

III I I II II I I III I I III

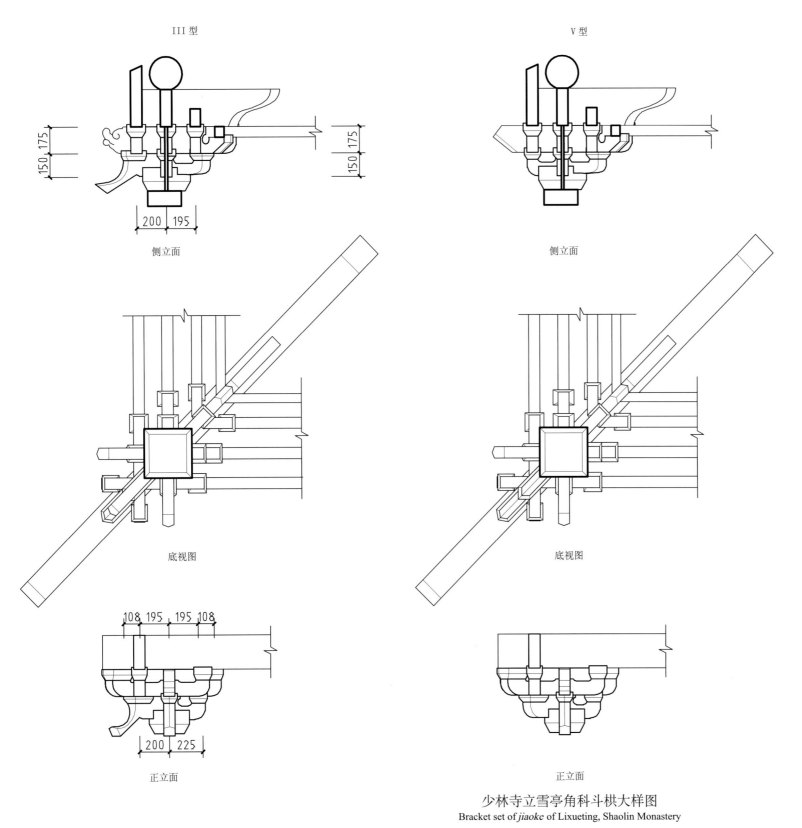

底视图

底视图

108 195 195 108

200 225

正立面

正立面

少林寺立雪亭角科斗栱大样图
Bracket set of *jiaoke* of Lixueting, Shaolin Monastery

0   0.5   1m

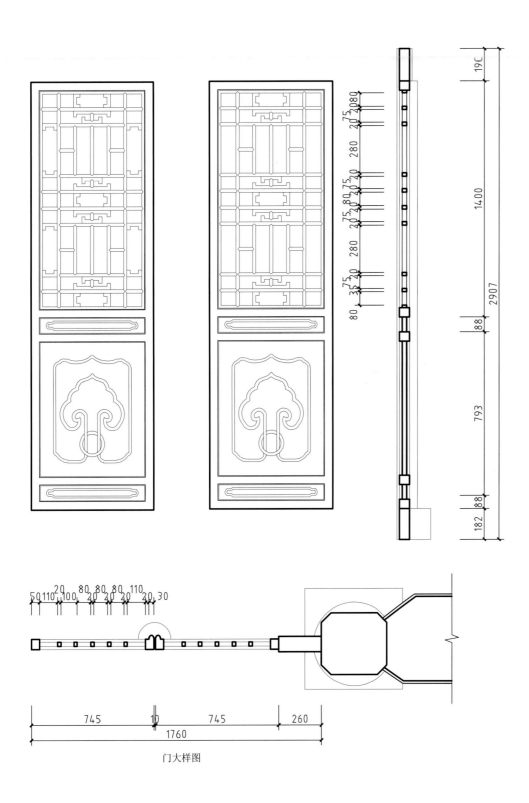

门大样图

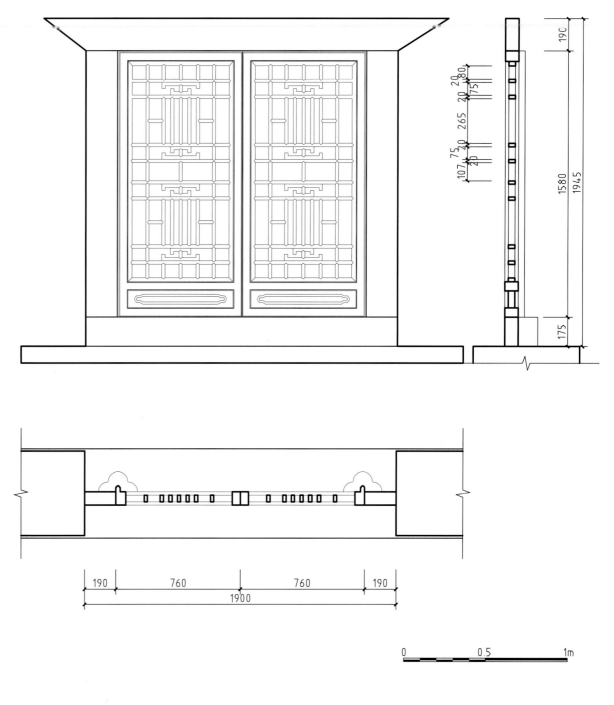

窗大样图

少林寺立雪亭门窗大样图
Doors and windows of Lixueting of Shaolin Monastery

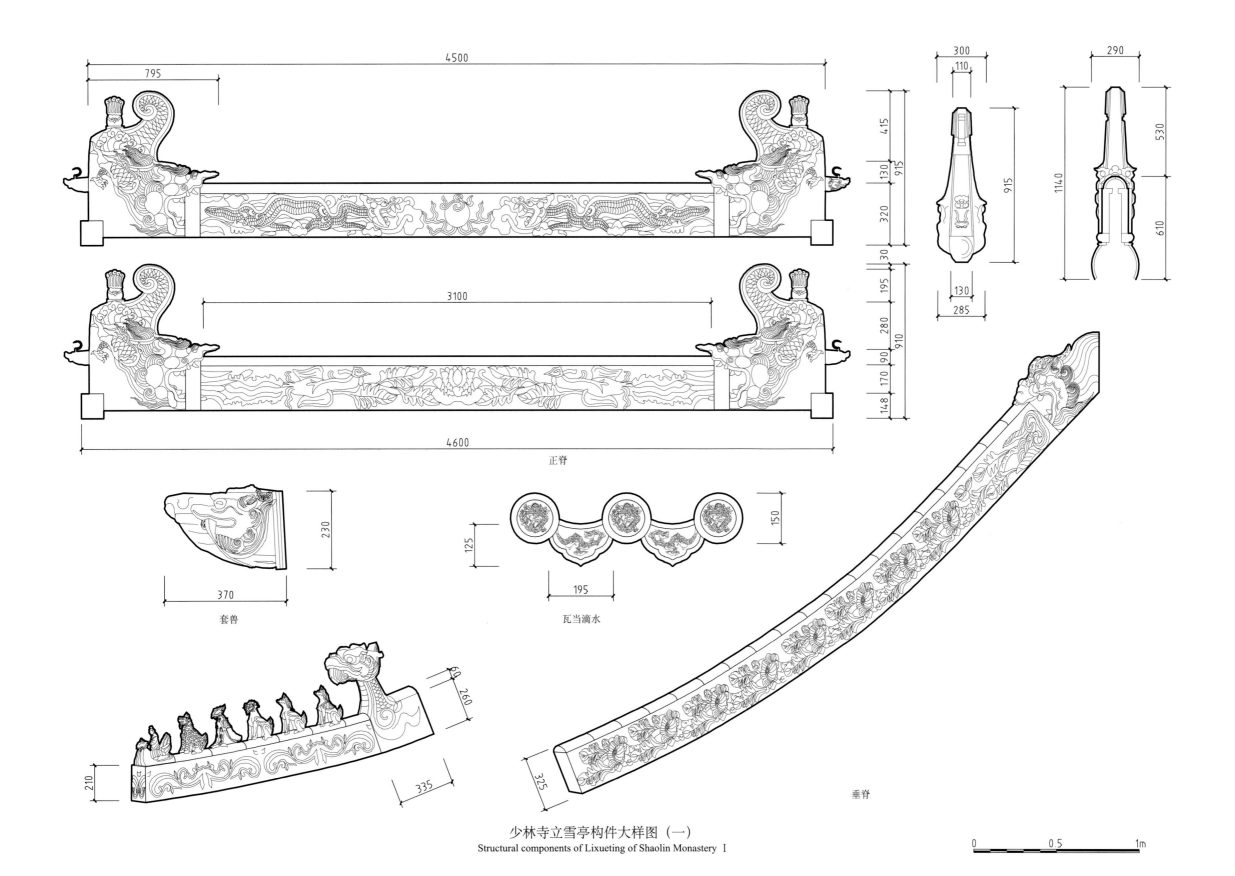

正脊

套兽

瓦当滴水

垂脊

少林寺立雪亭构件大样图（一）
Structural components of Lixueting of Shaolin Monastery Ⅰ

0   0.5   1m

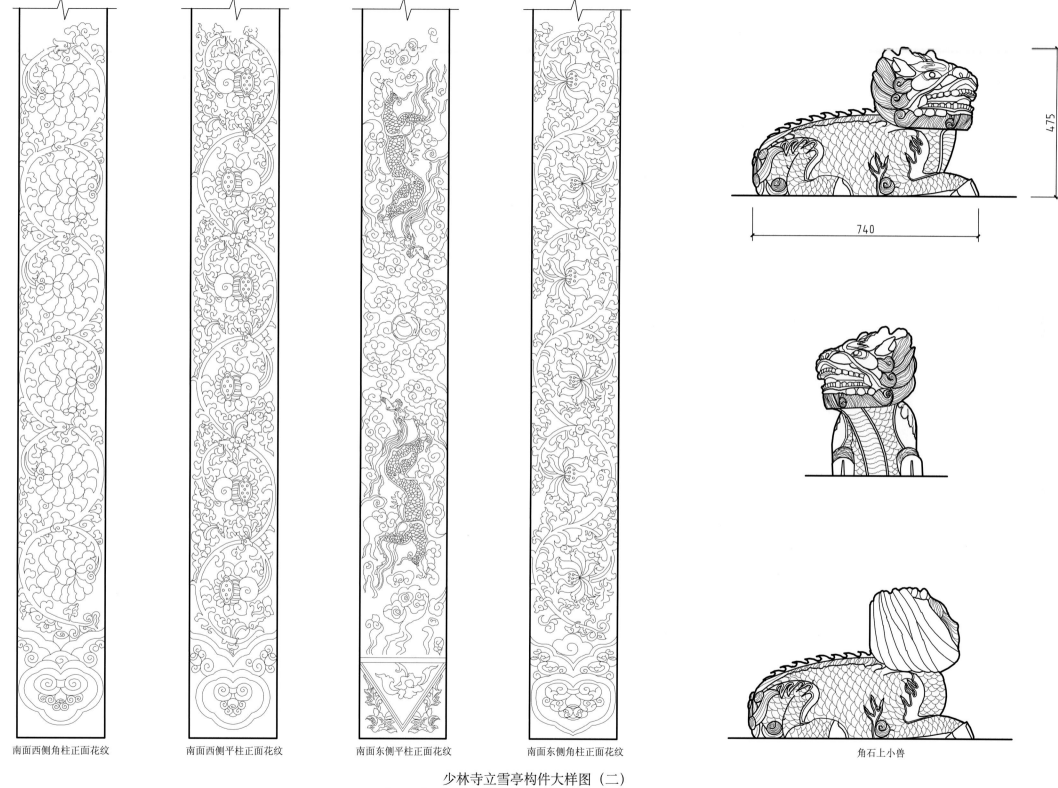

南面西侧角柱正面花纹　　　南面西侧平柱正面花纹　　　南面东侧平柱正面花纹　　　南面东侧角柱正面花纹

角石上小兽

少林寺立雪亭构件大样图（二）

Structural components of Lixueting of Lixueting of Shaolin Monastery Ⅱ

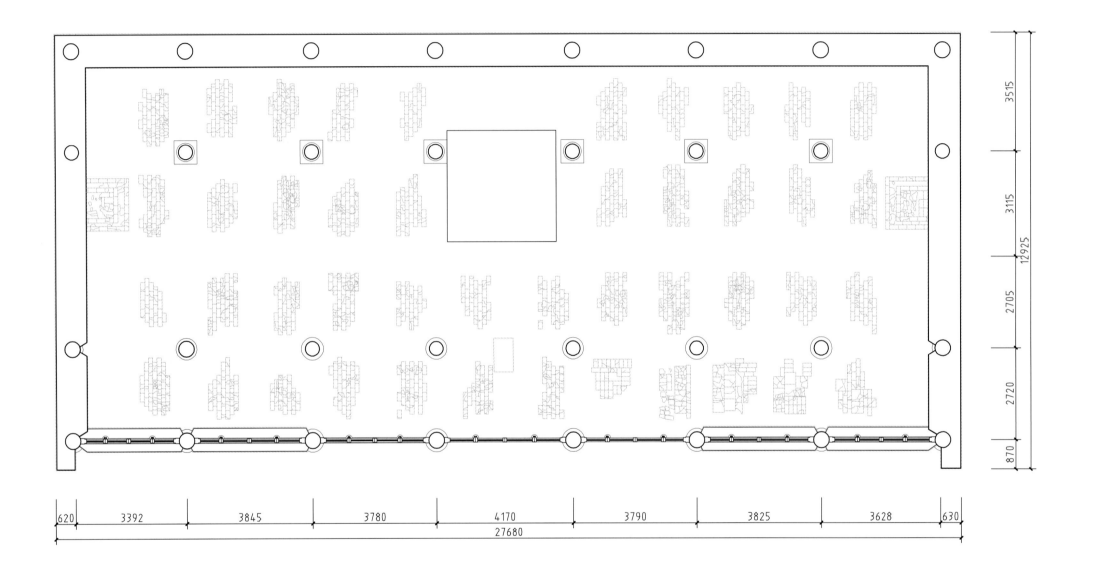

3515
3115
12925
2705
2720
870

620　3392　3845　3780　4170　3790　3825　3628　630
27680

N

0　1　2m

少林寺千佛殿平面图
Plan of Qianfodian of Lixueting of Shaolin Monastery

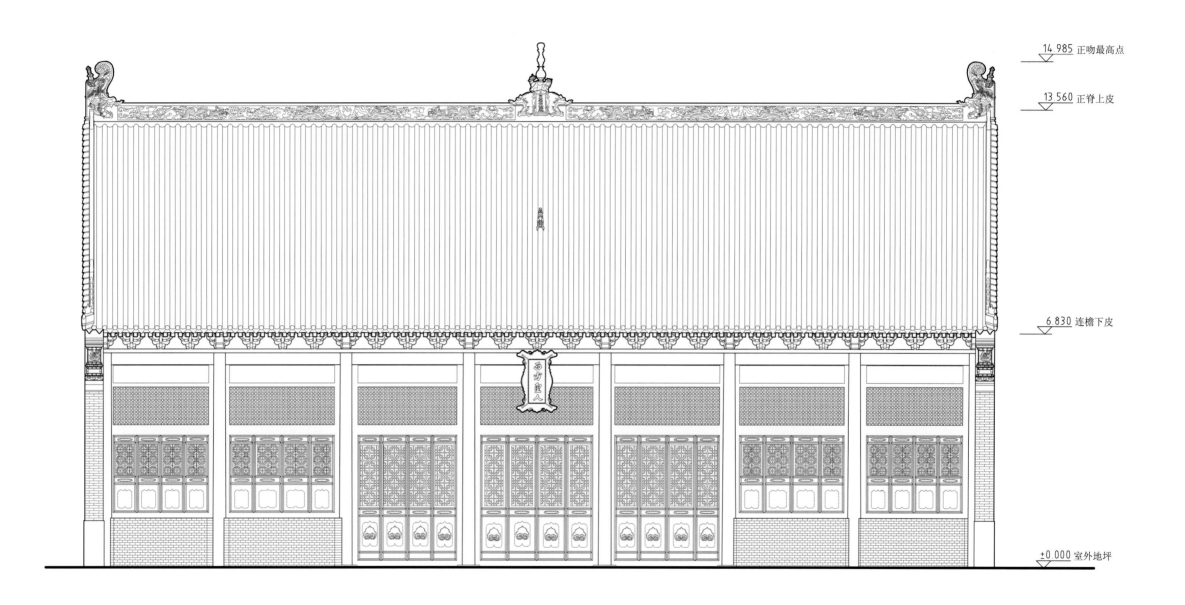

14.985 正吻最高点

13.560 正脊上皮

6.830 连檐下皮

±0.000 室外地坪

0 1 2m

少林寺千佛殿正立面图
Front elevation of Qianfodian of Lixueting of Shaolin Monastery

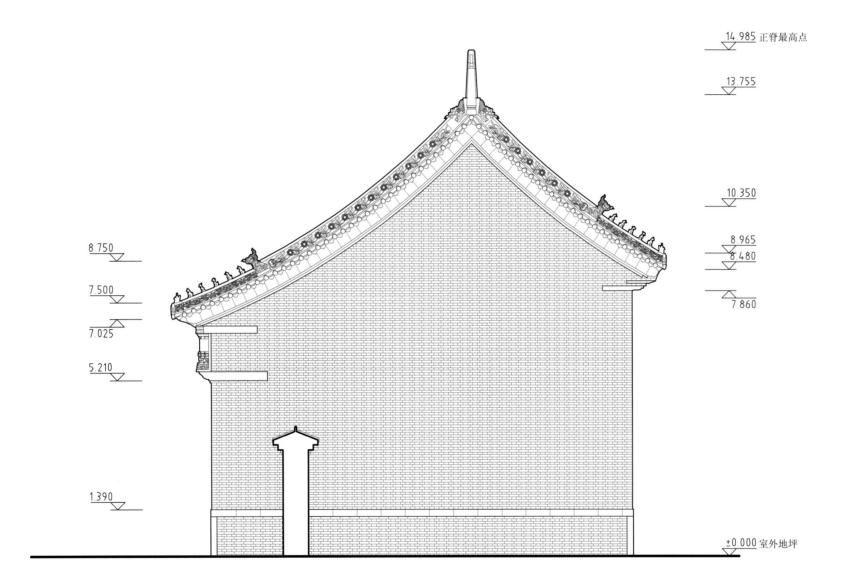

14.985 正脊最高点

13.755

10.350

8.965
8.480

7.860

8.750

7.500

7.025

5.210

1.390

±0.000 室外地坪

0 1 2m

少林寺千佛殿侧立面图
Side elevation of Qianfodian of Shaolin Monastery

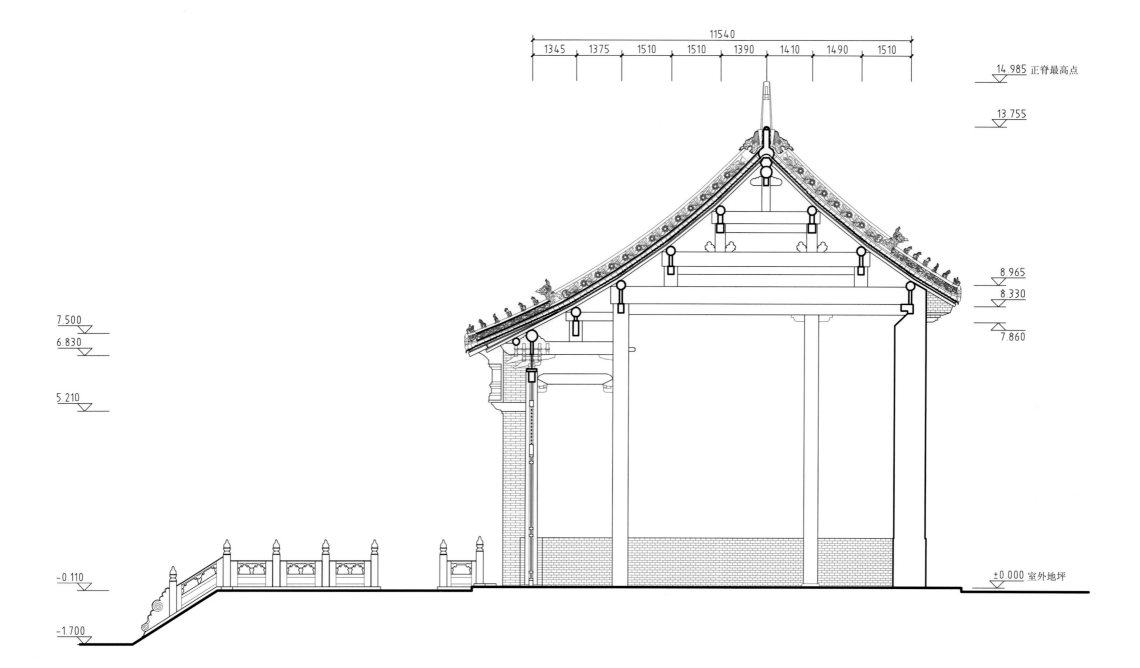

11540

1345　1375　1510　1510　1390　1410　1490　1510

14.985 正脊最高点

13.755

8.965

8.330

7.860

7.500

6.830

5.210

-0.110

-1.700

±0.000 室外地坪

少林寺千佛殿横剖面图

Cross-section of Qianfodian of Shaolin Monastery

0　1　2m

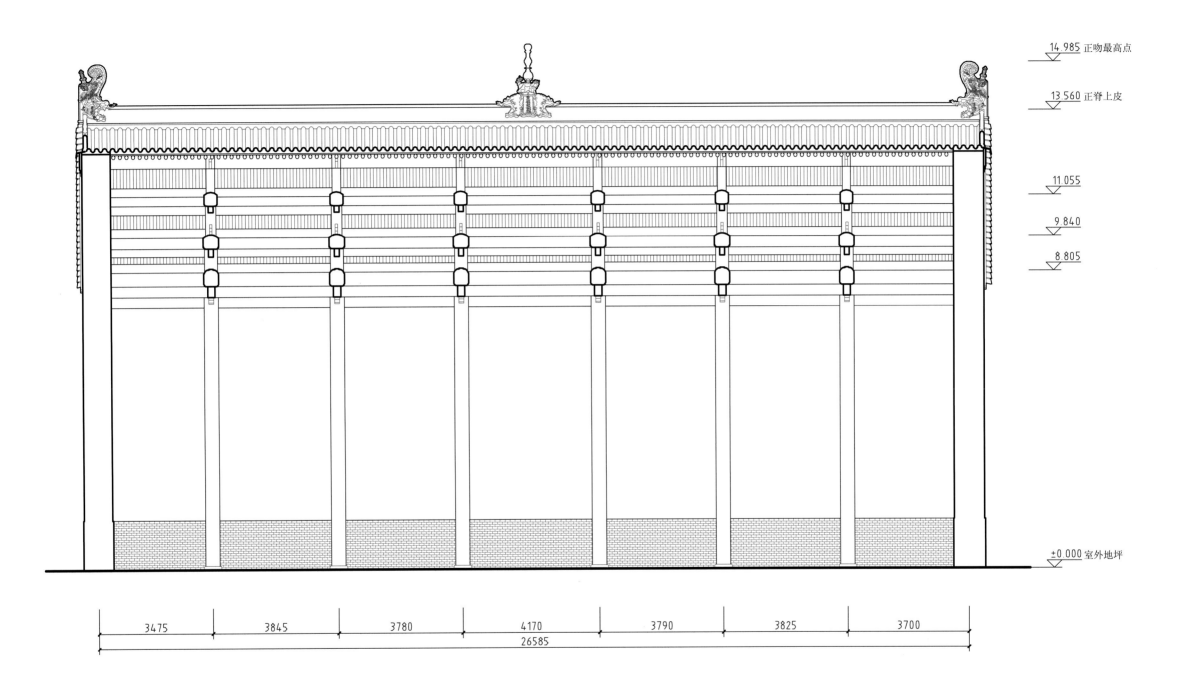

14.985 正吻最高点

13.560 正脊上皮

11.055

9.840

8.805

±0.000 室外地坪

| 3475 | 3845 | 3780 | 4170 | 3790 | 3825 | 3700 |

26585

少林寺千佛殿纵剖面图
Longitudinal section of Qianfodian of Shaolin Monastery

0 1 2m

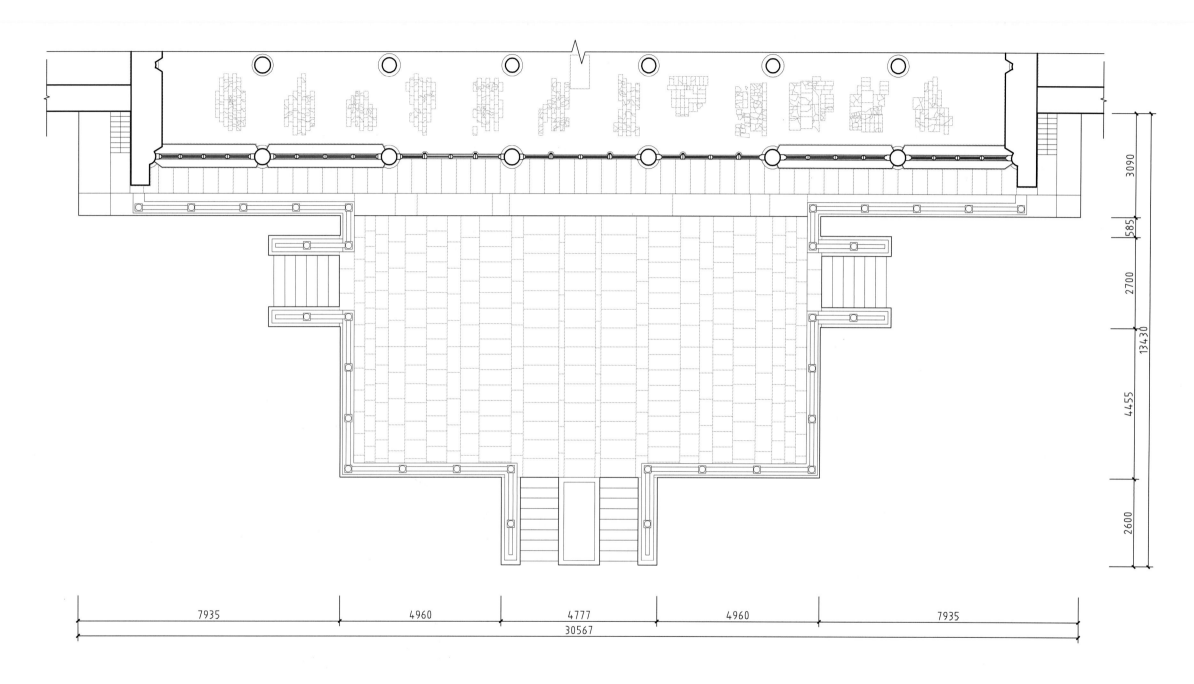

N

0 1 2m

少林寺千佛殿台基平面图
Plan of *taiji* of Qianfodian, Shaolin Monastery

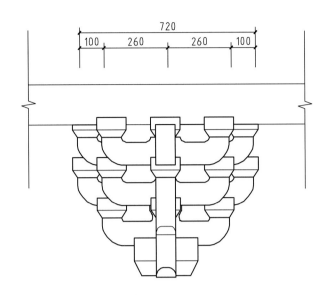

少林寺千佛殿平身科斗栱大样图
Bracketing of *pingshenke* of Qianfodian, Shaolin Monastery

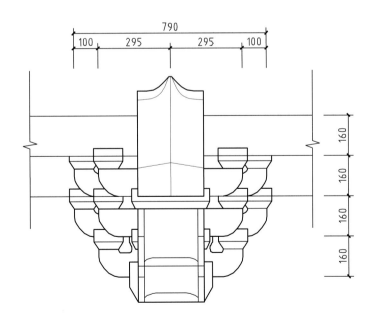

少林寺千佛殿柱头科斗栱大样图
Bracketing of *zhutouke* of Qianfodian, Shaolin Monastery

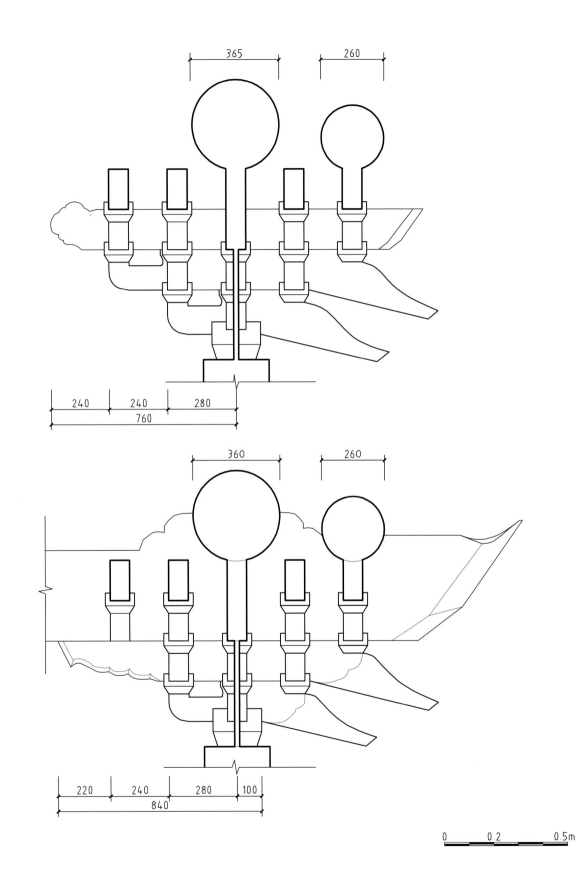

0   0.2   0.5m

少林寺千佛殿墀头大样图
Chitou of Qianfodian of Shaolin Monastery

0   0.1 0.2m

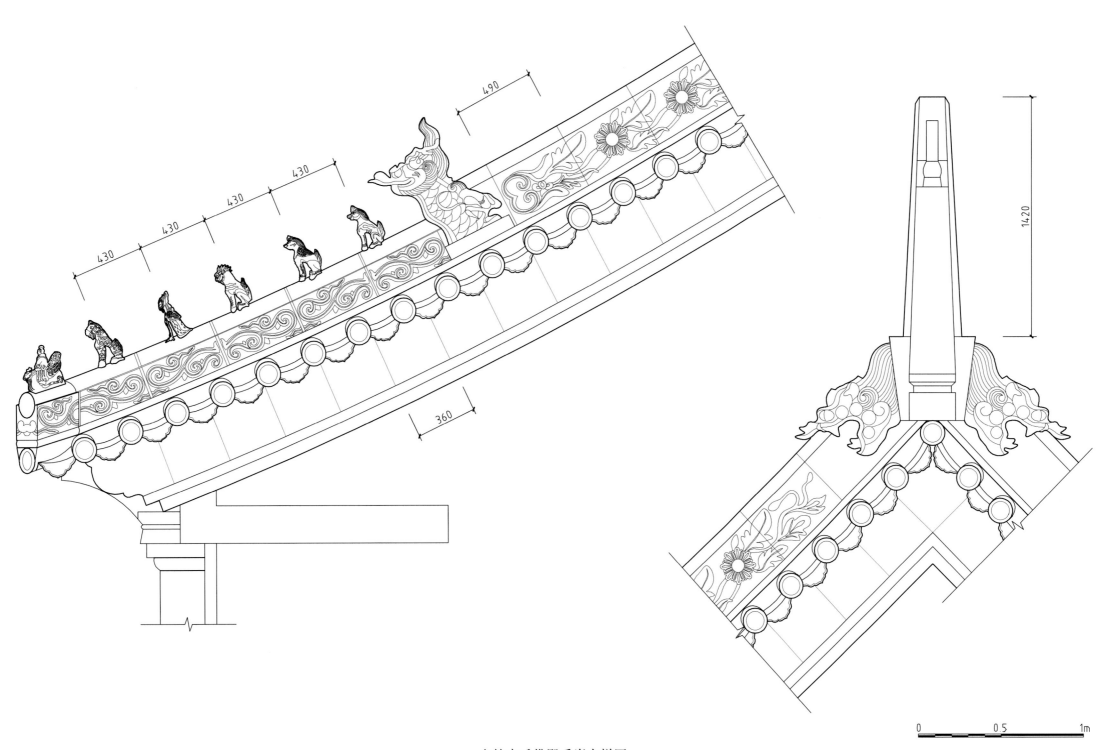

少林寺千佛殿垂脊大样图
Chuiji of Qianfodian of Shaolin Monastery

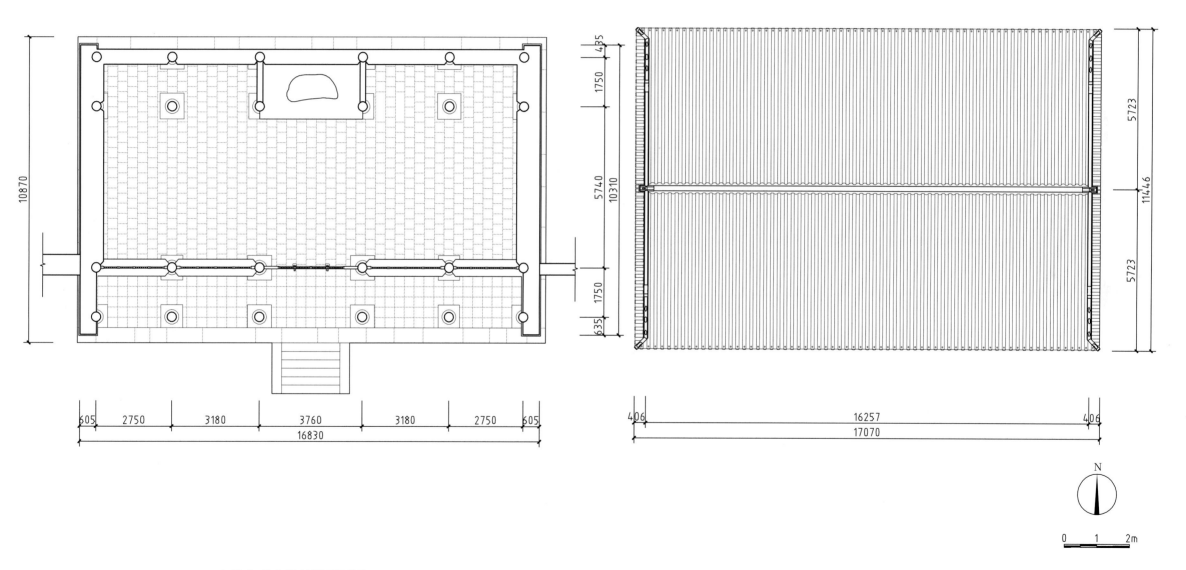

少林寺观音殿首层平面图
Plan of first floor of Guanyindian, Shaolin Monastery

少林寺观音殿屋顶平面图
Roof plan of Guanyindian of Shaolin Monastery

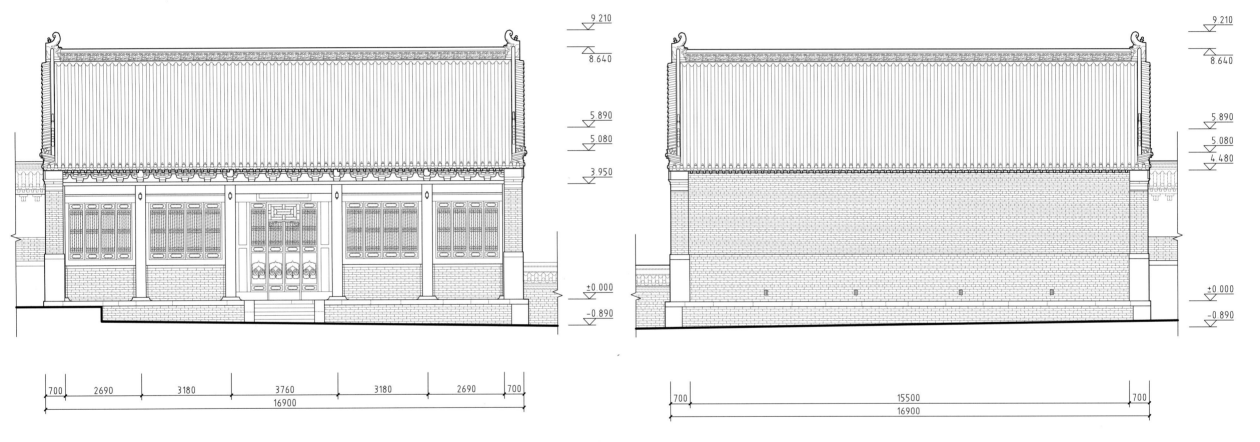

少林寺观音殿正立面图
Front elevation of Guanyindian of Shaolin Monastery

少林寺观音殿背立面图
Rear elevation of Guanyindian of Shaolin Monastery

0　1　2m

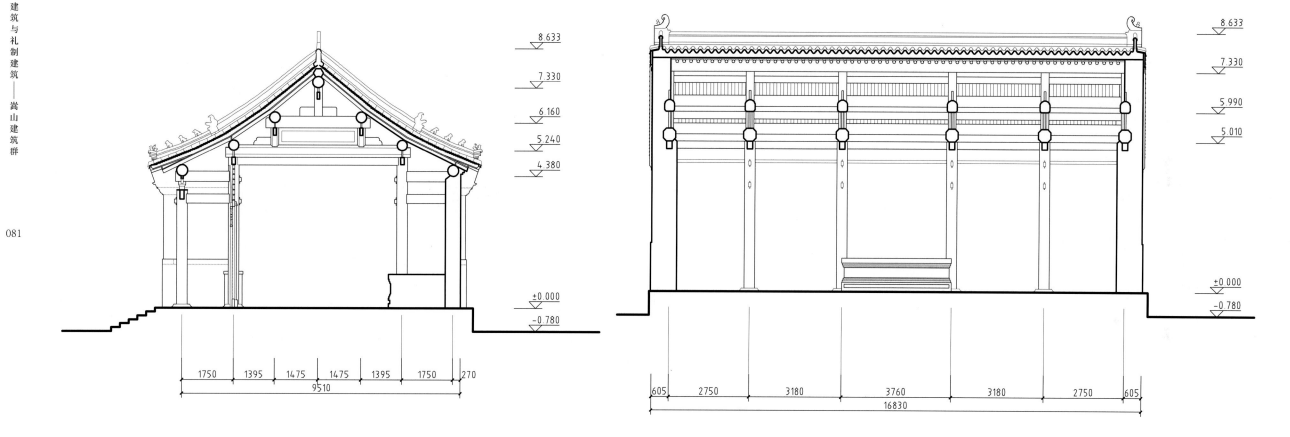

8.633
7.330
6.160
5.240
4.380
±0.000
-0.780

1750　1395　1475　1475　1395　1750　270
9510

8.633
7.330
5.990
5.010
±0.000
-0.780

605　2750　3180　3760　3180　2750　605
16830

少林寺观音殿横剖面图
Cross-section of Guanyindian of Shaolin Monastery

少林寺观音殿纵剖面图
Longitudinal section of Guanyindian of Shaolin Monastery

N

0　1　2m

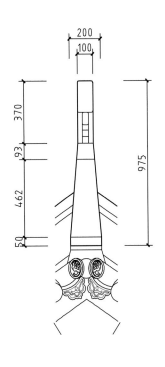

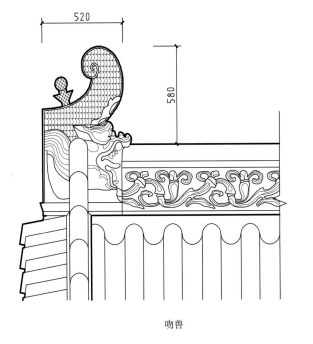

吻兽

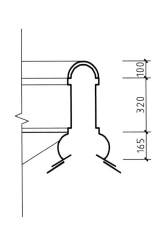

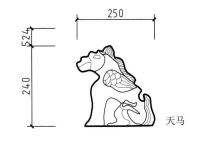

天马

狮

龙

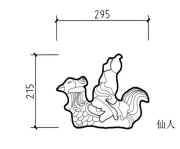

仙人

仙人走兽

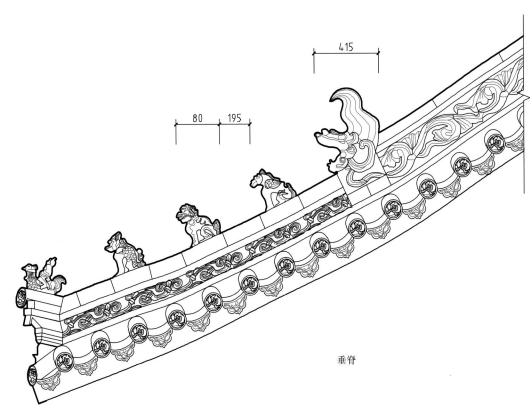

垂脊

少林寺观音殿构件大样图
Structural components of Guanyindian of Shaolin Monastery

0　　　　　0.5　　　　　1m

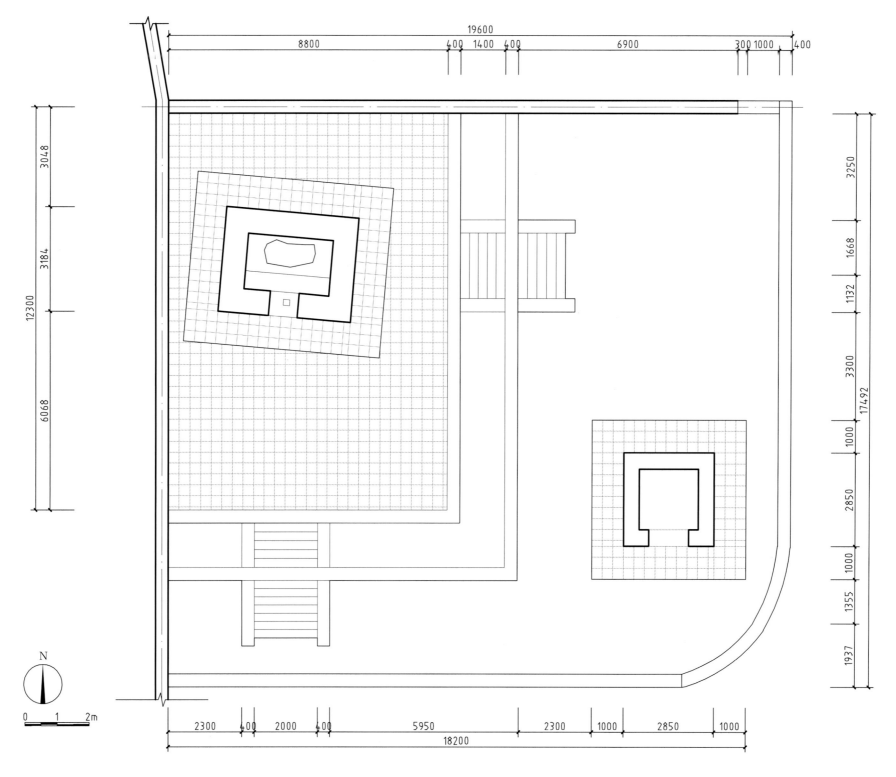

少林寺塔院总平面图
Site plan of Tayuan of Shaolin Monastery

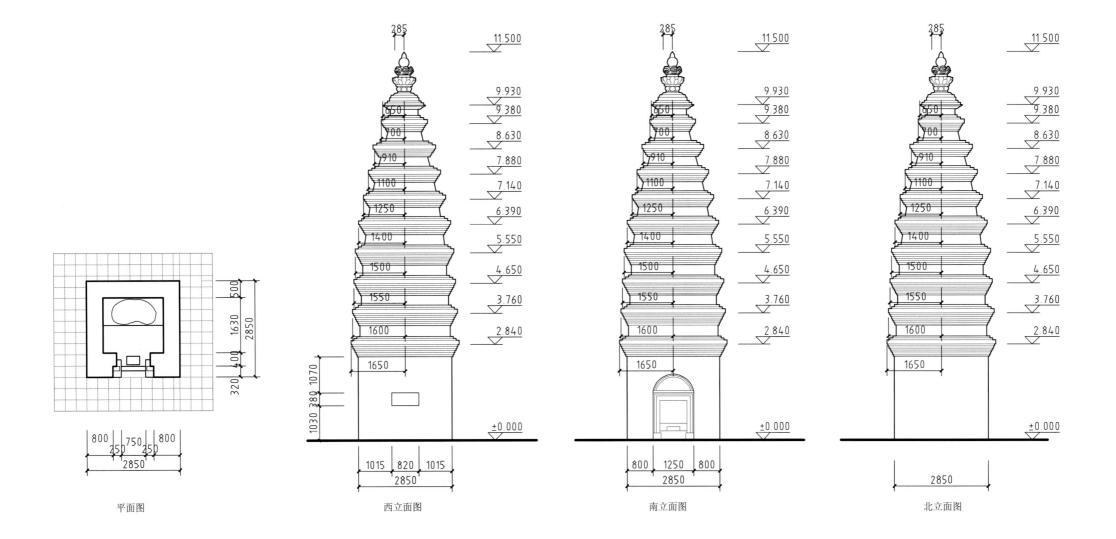

平面图　　　　　　　西立面图　　　　　　南立面图　　　　　　北立面图

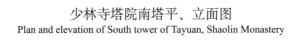

少林寺塔院南塔平、立面图
Plan and elevation of South tower of Tayuan, Shaolin Monastery

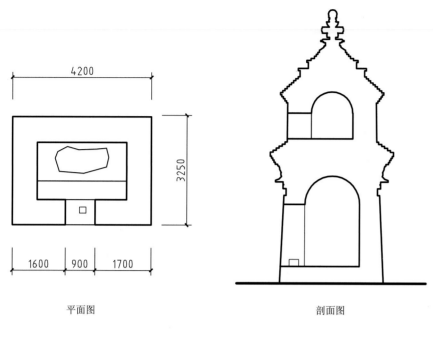

4200

3250

1600 900 1700

平面图

剖面图

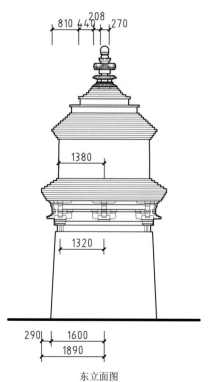

810 440 208 270

1380

1320

290 1600
1890

东立面图

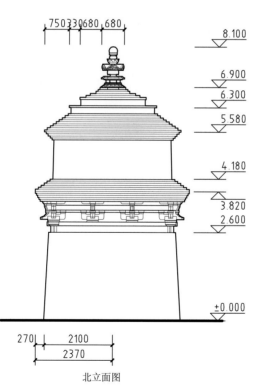

750 330 680 680

8.100

6.900
6.300

5.580

4.180

3.820
2.600

±0.000

270 2100
2370

北立面图

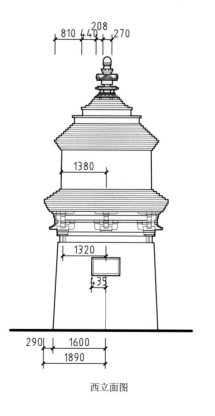

810 440 208 270

1380

1320

435

290 1600
1890

西立面图

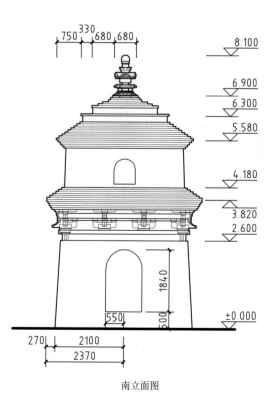

330 680 680

750

8.100

6.900
6.300

5.580

4.180

3.820
2.600

1840

550 500

±0.000

270 2100
2370

南立面图

少林寺塔院北塔平、立、剖面图
Plan, elevation and section of North tower of Tayuan, Shaolin Monastery

0 1 2m

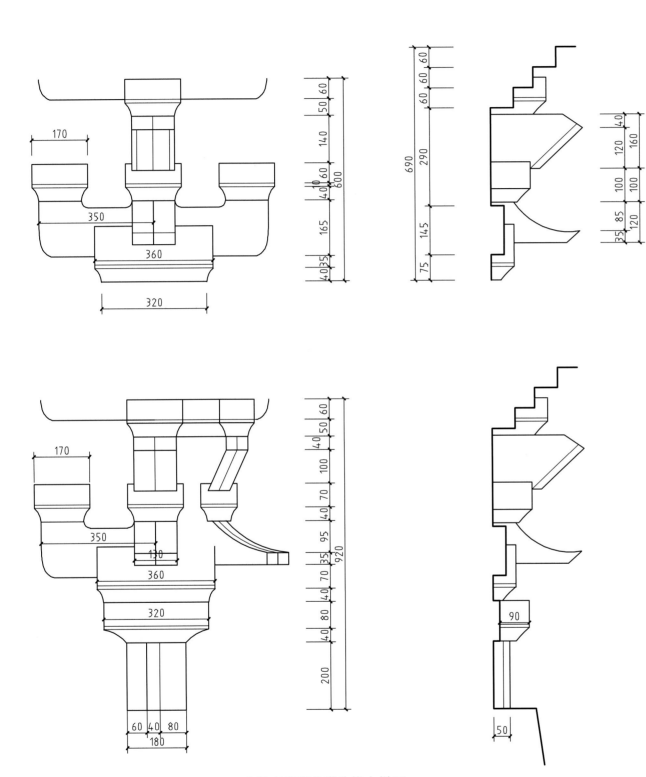

少林寺塔院北塔斗栱大样图
Bracket set of North tower of Tayuan, Shaolin Monastery

少林寺塔院北塔塔刹大样图
Tasha of North tower of Tayuan, Shaolin Monastery

0   0.1   0.2m

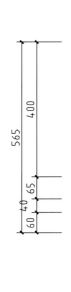

少林寺塔院南塔塔门大样图
Entrance of South tower of Tayuan, Shaolin Monastery

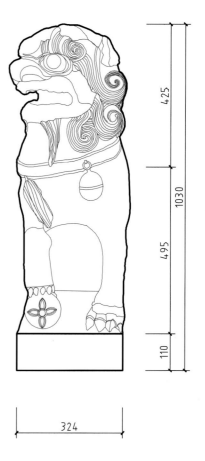

少林寺塔院南塔石狮大样图
Stone lion of South tower of Tayuan, Shaolin Monastery

0　　0.1　0.2m

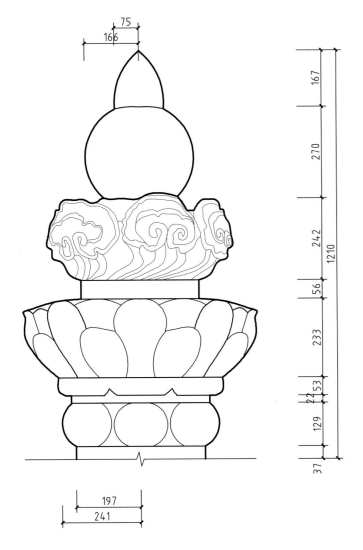

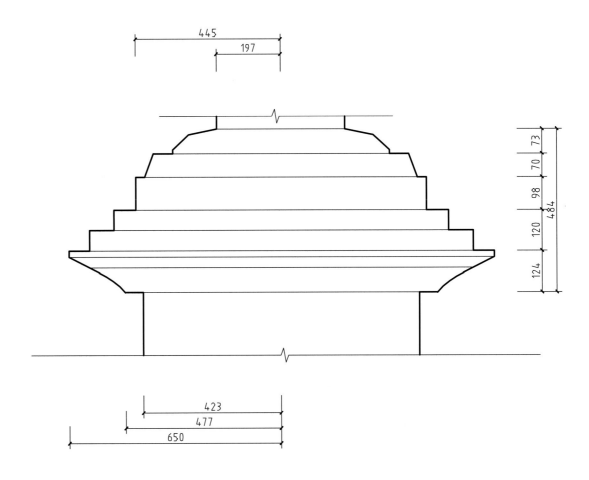

少林寺塔院南塔塔刹及基座大样图
Tasha and *jizuo* of South tower of Tayuan, Shaolin Monastery

0　0.1　0.2m

坊以吾佛
釋迦 渚经藏廣說
彌勒世尊見居知品天宮
化四十九重内院五伯萬
億 宫嘉 殊祥莫可稱
數畫夜懷声不聞有苦海
不退法救渚天子禍盡退
墮三途惡道
佛勸僧俗仁者須 信向
精勤 供卷念名 觀
事法終二回之
往生蓮問上品棉奉尊記
尚天快樂隨
佛下生龍 初會授菩提
記廣慶建慈佛塔准一
切衆生
見聞信喜省悟終行止世
它生同成心覺滿四誓
生功德上祝 人
少帝萬歲
太皇太后 皇太后
聖壽延遠安樂吉祥國泰
民康榮生無惱之佑丁卯
中 少林住山廣慶謹

少林寺塔院北塔侧面碑文大样图

Side inscription on North tower of Tayuan, Shaolin Monastery

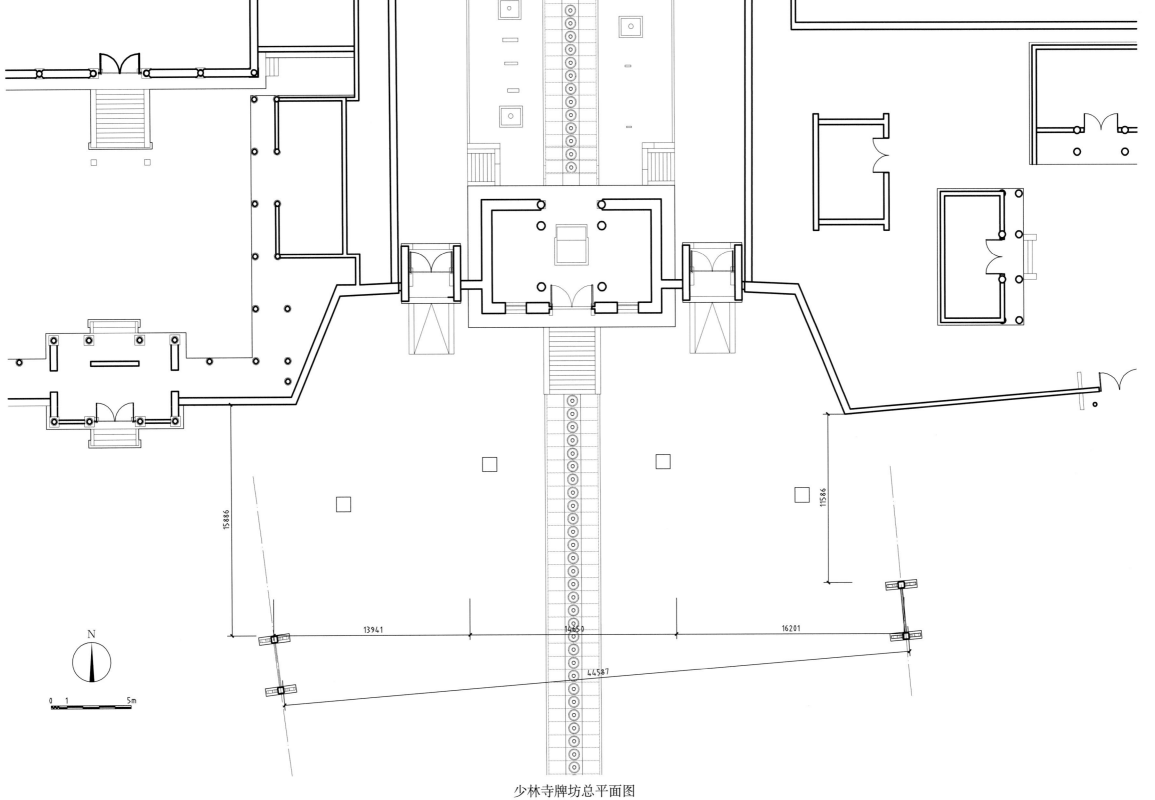

N

0 1   5m

少林寺牌坊总平面图
Site plan of *paifang* of Shaolin Monastery

少林寺西侧牌坊西立面图

West elevation of west-side *paifang* of Shaolin Monastery

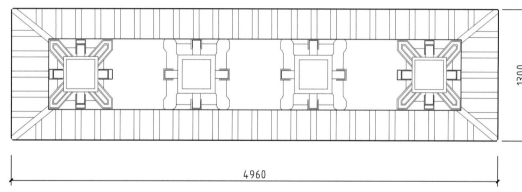

少林寺西侧牌坊仰视平面图

Plan of west-side *paifang* of Shaolin Monastery as seen from below

0　0.5　1m

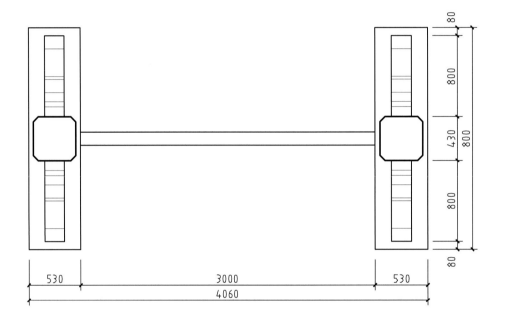

少林寺西侧牌坊平面图
Plan of west-side *paifang* of Shaolin Monastery

少林寺西侧牌坊东立面图
East elevation of west-side *paifang* of Shaolin Monastery

0　　0.5　　1m

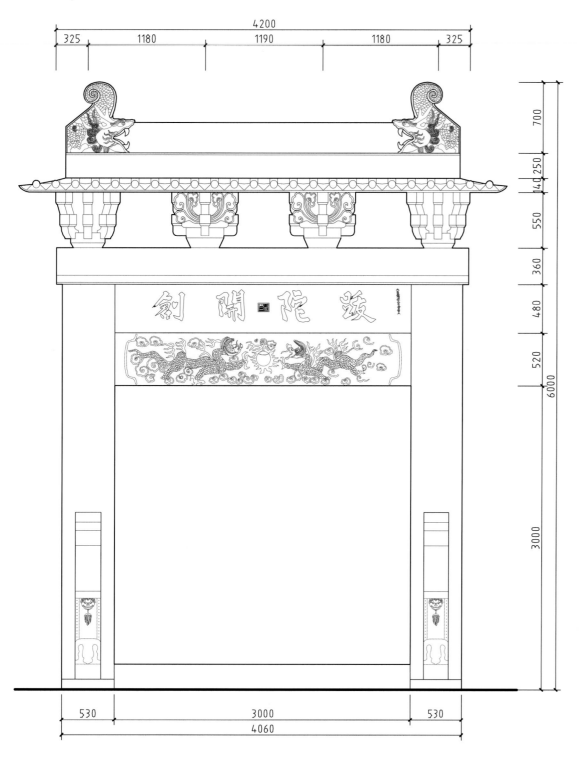

少林寺东侧牌坊西立面图

West elevation of east-side *paifang* of Shaolin Monastery

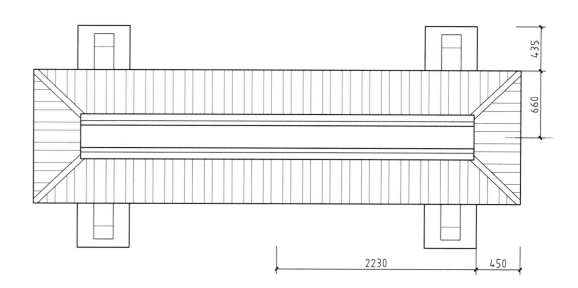

少林寺东侧牌坊屋顶平面图

Roof plan of east-side *paifang* of Shaolin Monastery

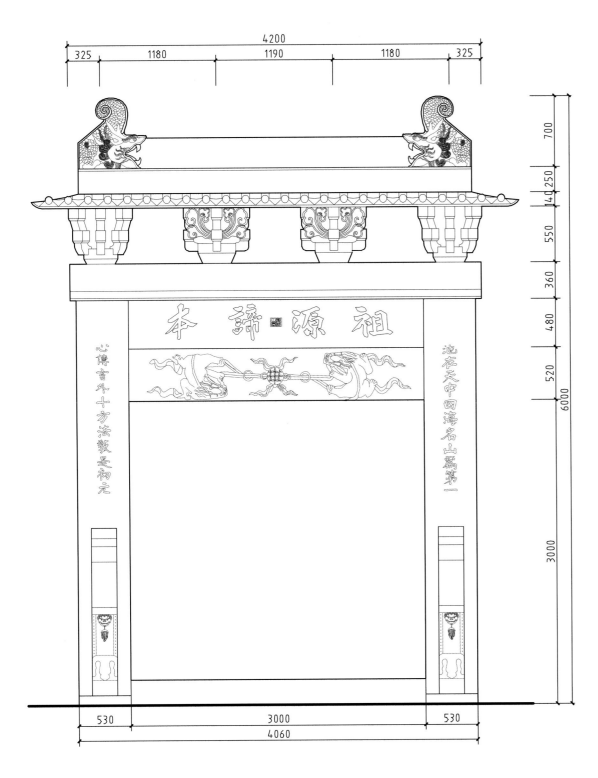

少林寺东侧牌坊东立面图

East elevation of east-side *paifang* of Shaolin Monastery

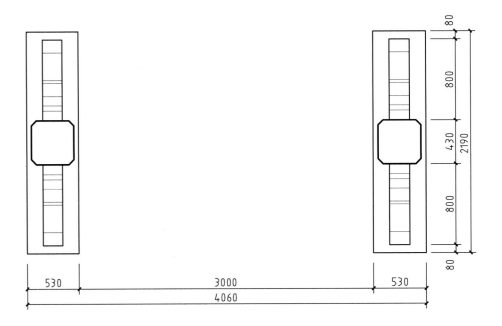

少林寺东侧牌坊平面图

Plan of east-side *paifang* of Shaolin Monastery

0　0.5　1m

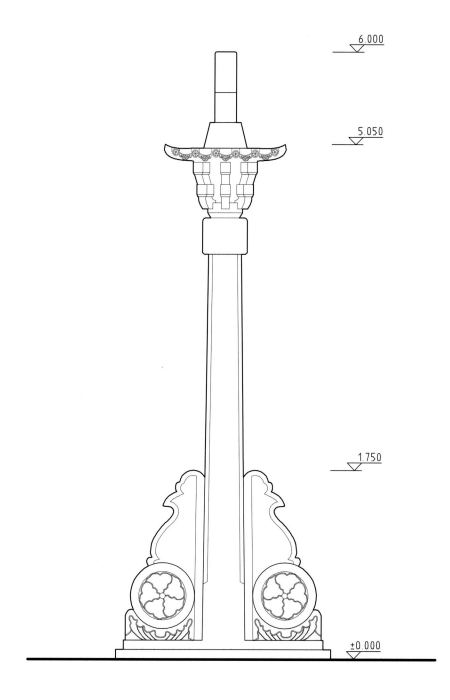

6.000

5.050

1.750

±0.000

少林寺山门前牌坊侧立面图
Side elevation of frontal paifang of *shanmen*, Shaolin Monastery

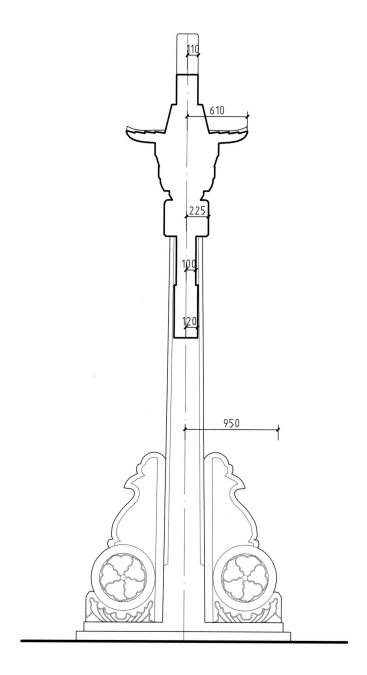

110

610

225

100

120

950

少林寺山门前牌坊剖面图
Section of frontal paifang of *shanmen*, Shaolin Monastery

0   0.5   1m

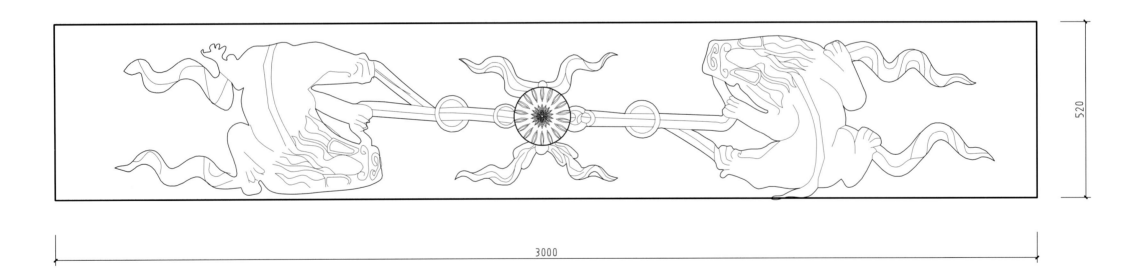

少林寺牌坊额枋纹样大样图（一）

Decoration of *efang* of *paifang*, Shaolin Monastery Ⅰ

0　0.1　0.2m

520

3000

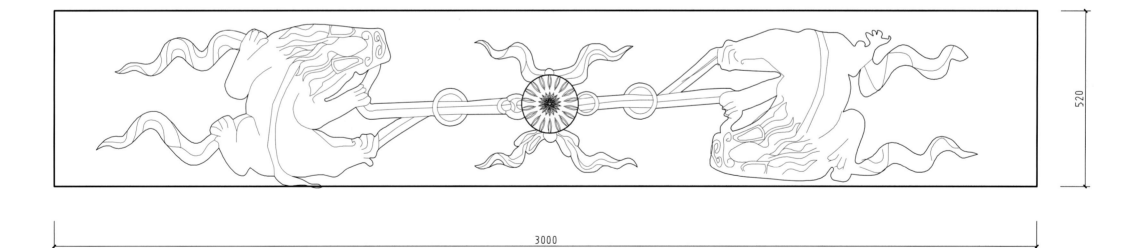

520

3000

少林寺牌坊额枋纹样大样图（二）
Decoration of *efang* of *paifang*, Shaolin Monastery Ⅱ

0 0.1 0.2m

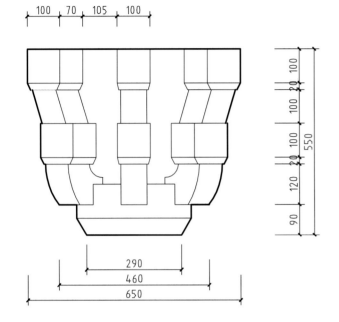

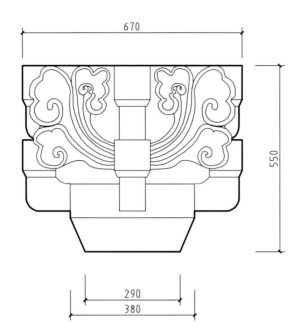

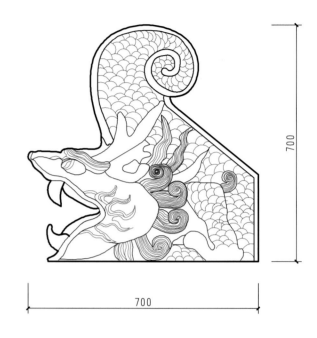

少林寺牌坊鸱尾大样图
Chiwei of *paifang* of Shaolin Monastery

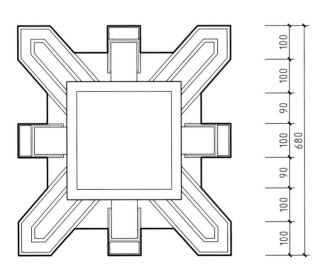

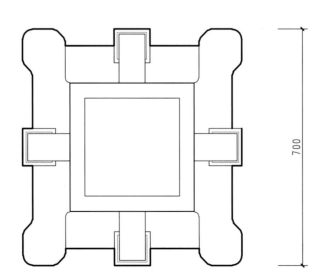

少林寺牌坊斗栱大样图
Bracket set of *paifang* of Shaolin Monastery

0   0.1   0.2m

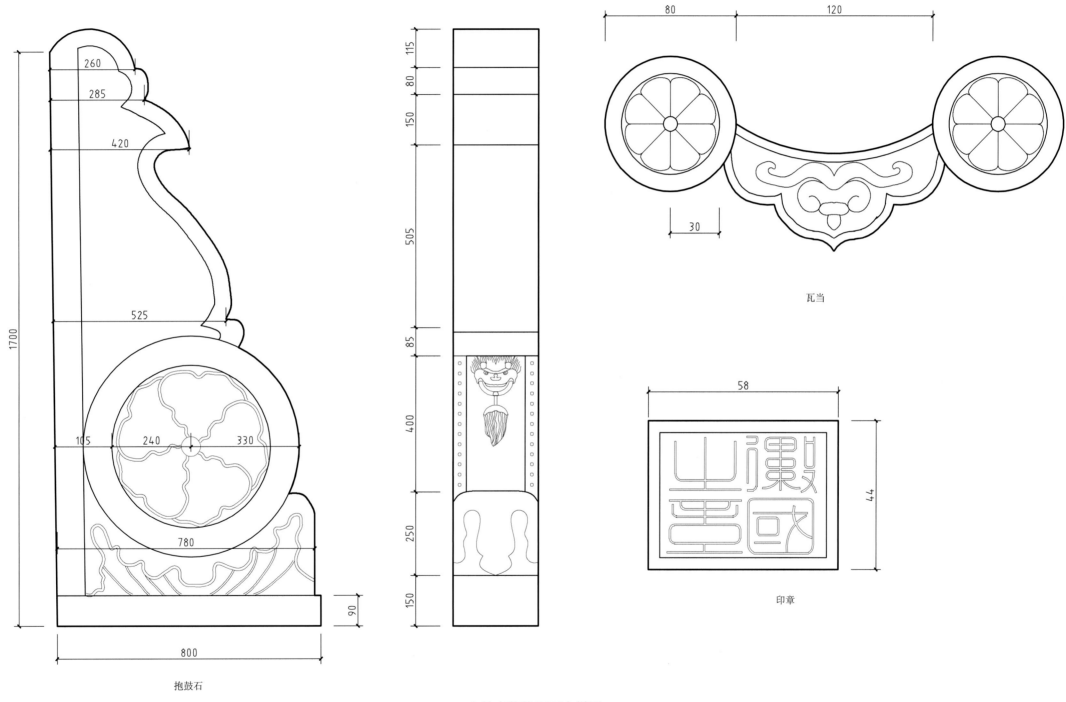

抱鼓石

瓦当

印章

少林寺牌坊瓦石大样图
Tiles of *paifang* of Shaolin Monastery

0 0.1 0.2m

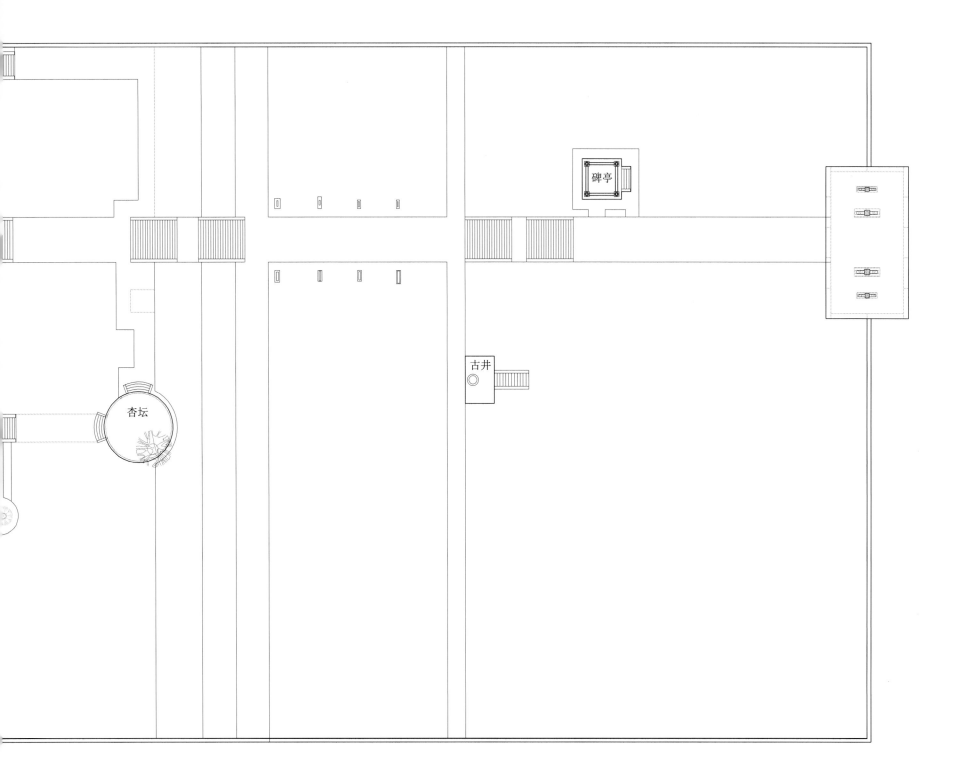

嵩阳书院
Songyang Academy

碑亭

古井

杏坛

0   5   20m

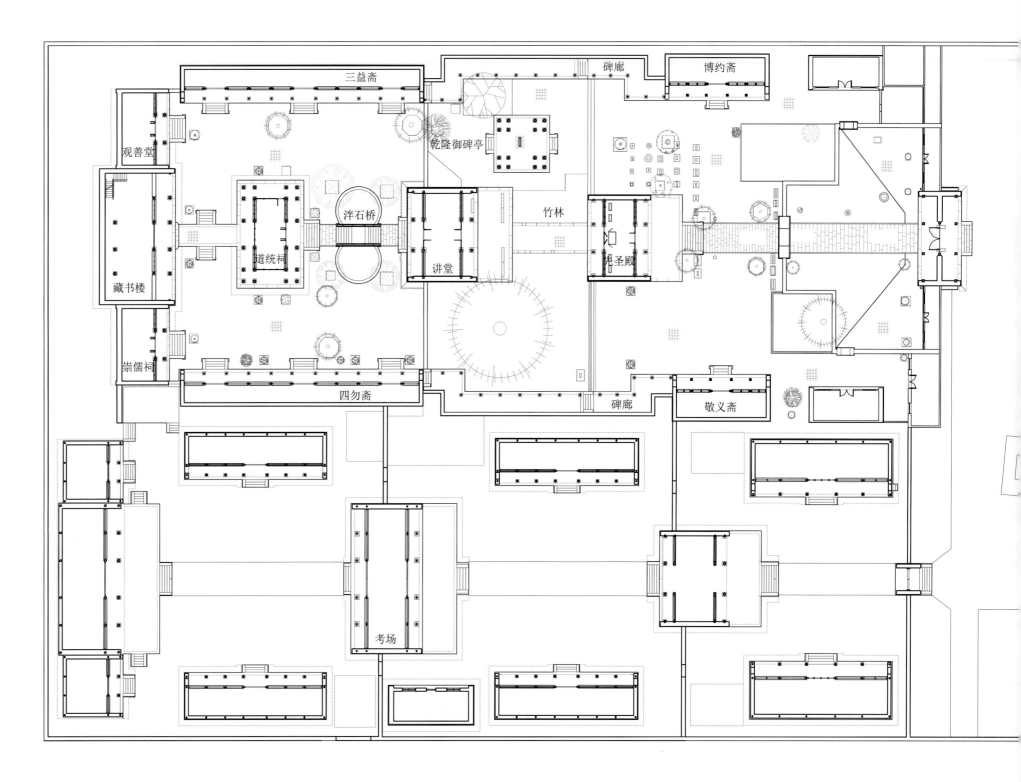

三益斋

碑廊

博约斋

观善堂

乾隆御碑亭

泮石桥

竹林

道统祠

讲堂

先圣殿

藏书楼

崇儒祠

四勿斋

碑廊

敬义斋

考场

嵩阳书院总平面图
Site plan of Songyang Academy

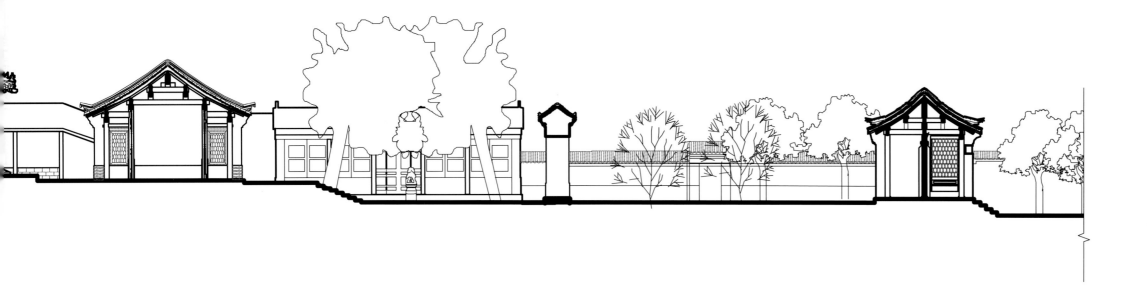

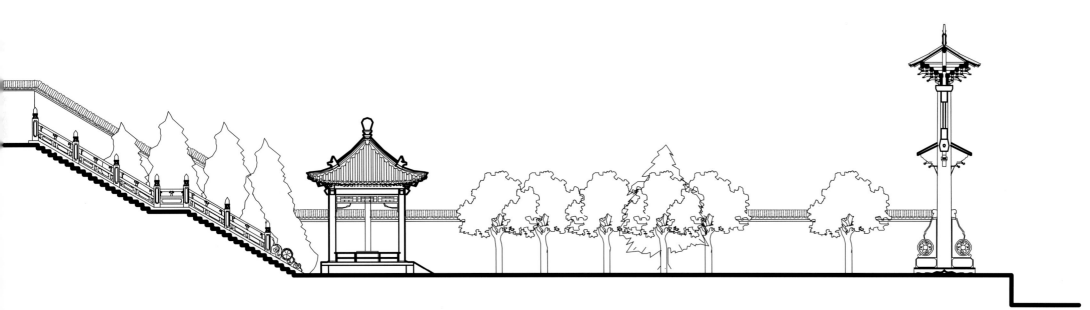

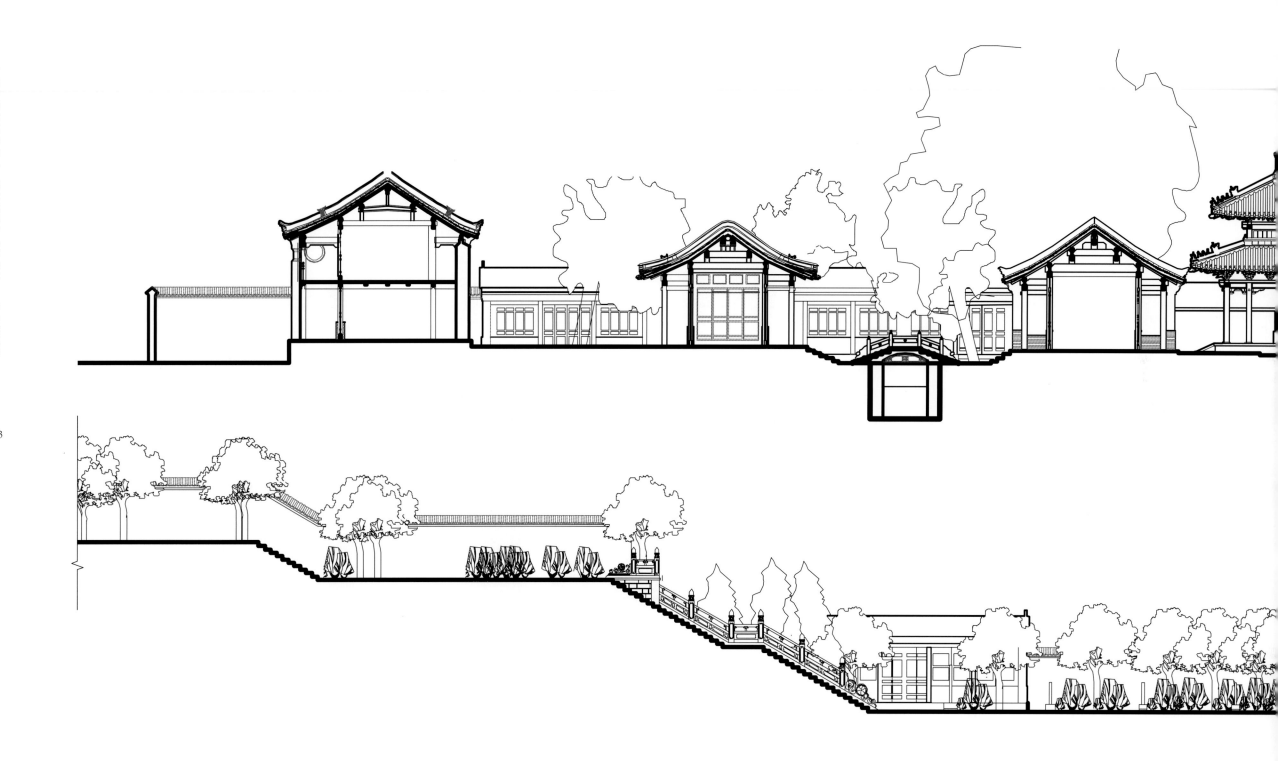

嵩阳书院总纵剖面图
Longitudinal section of Songyang Academy

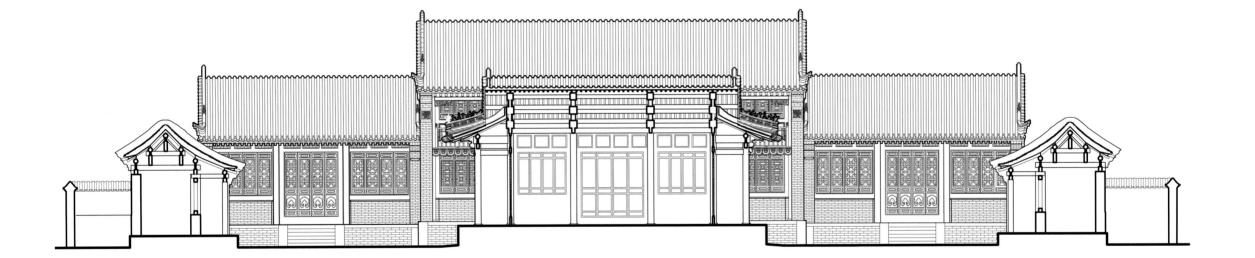

嵩阳书院总横剖面图
Cross-section of Songyang Academy

0 1 2m

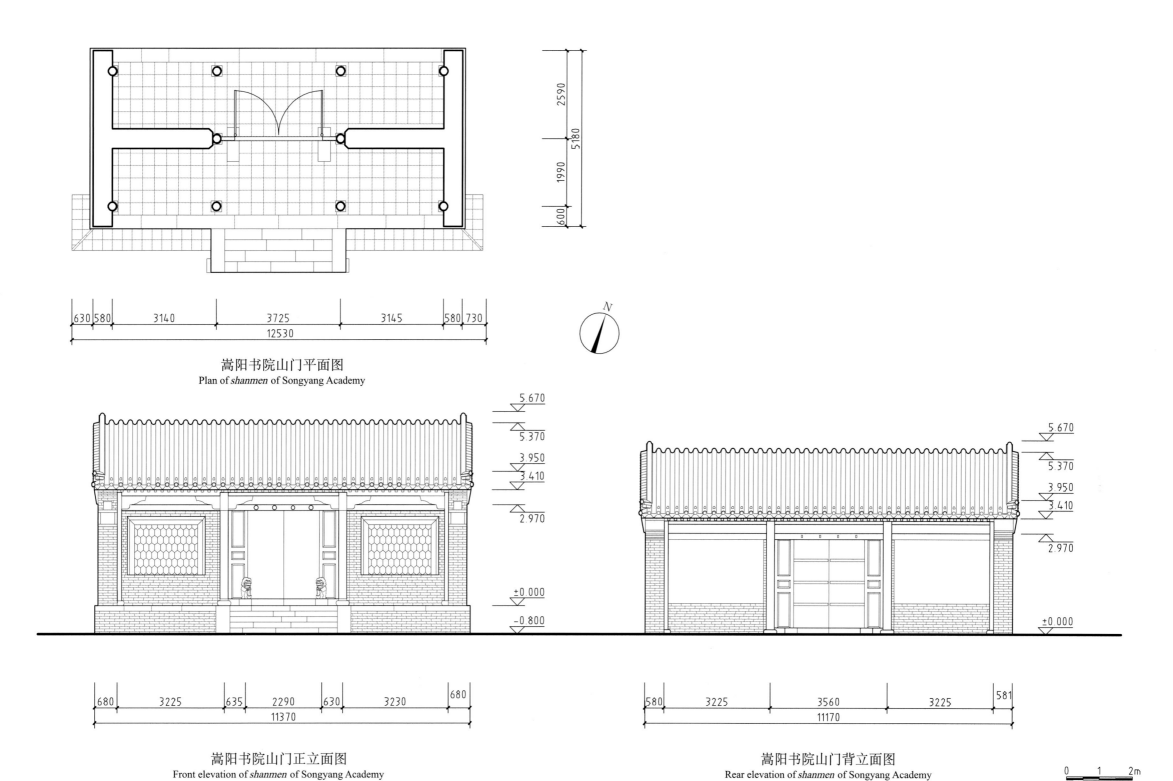

嵩阳书院山门平面图
Plan of *shanmen* of Songyang Academy

嵩阳书院山门正立面图
Front elevation of *shanmen* of Songyang Academy

嵩阳书院山门背立面图
Rear elevation of *shanmen* of Songyang Academy

0 1 2m

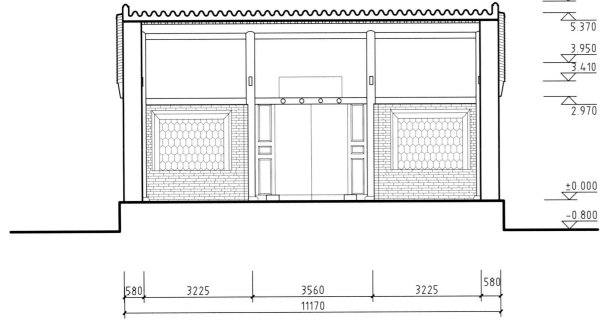

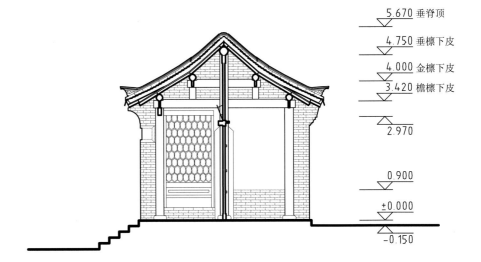

嵩阳书院山门纵剖面图
Longitudinal section of *shanmen* of Songyang Academy

嵩阳书院山门横剖面图
Cross-section of *shanmen* of Songyang Academy

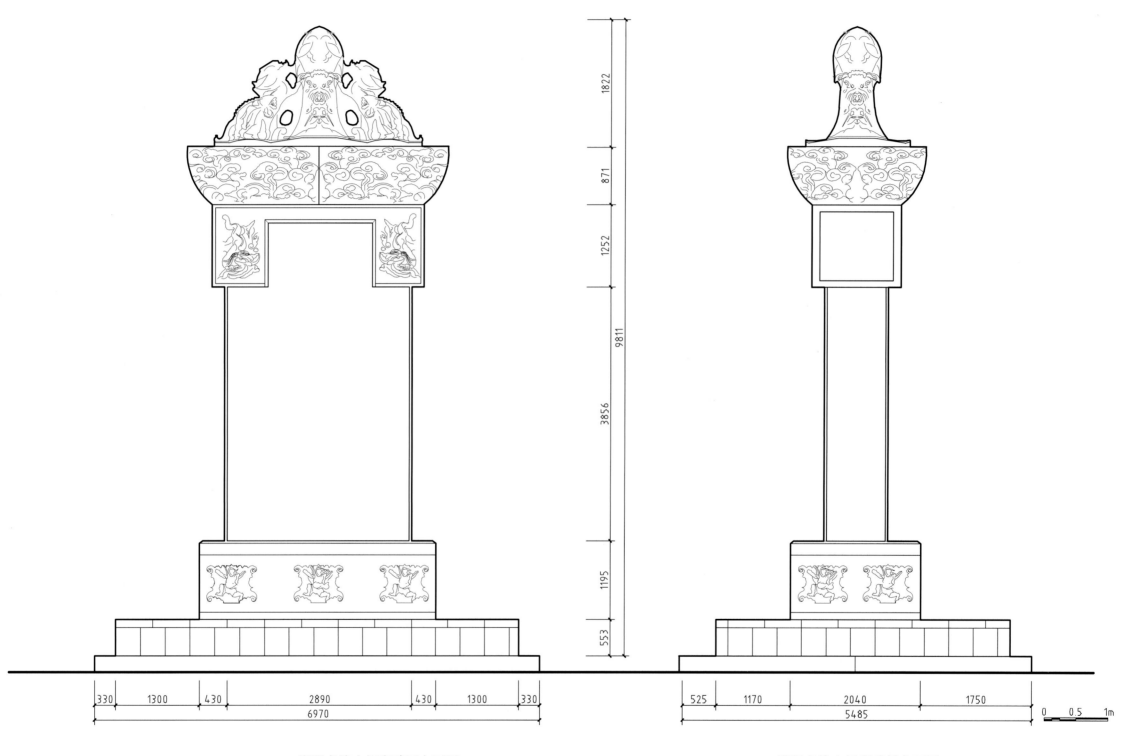

嵩阳书院山门唐碑正立面图
Front elevation of Tangbei of *shanmen*, Songyang Academy

嵩阳书院山门唐碑侧立面图
Side elevation of Tangbei of *shanmen*, Songyang Academy

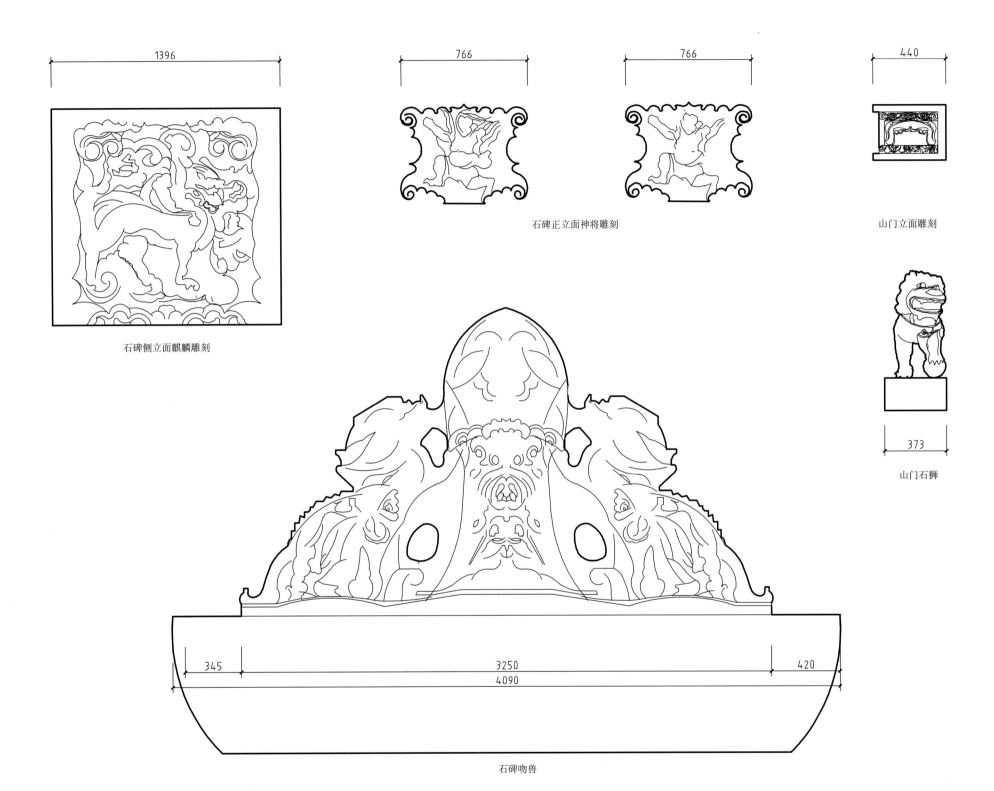

1396

766

766

440

石碑侧立面麒麟雕刻

石碑正立面神将雕刻

山门立面雕刻

373

山门石狮

345

3250

420

4090

石碑吻兽

嵩阳书院山门唐碑构件大样图
Structural components of Tangbei of *shanmen*, Songyang Academy

0   0.5   1m

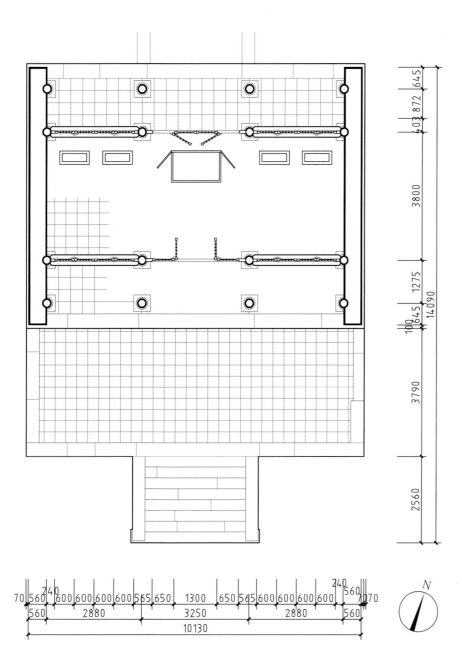

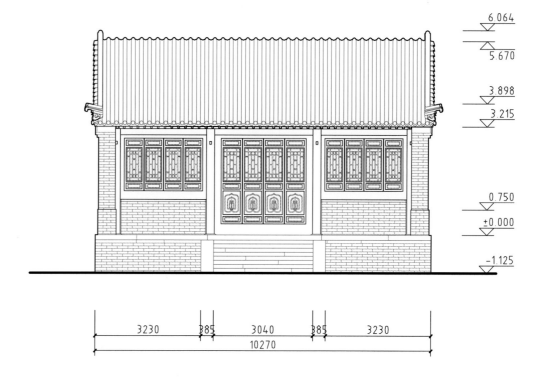

嵩阳书院先圣殿平面图
Plan of Xianshengdian of Songyang Academy

嵩阳书院先圣殿正立面图
Front elevation of Xianshengdian of Songyang Academy

0   1   2m

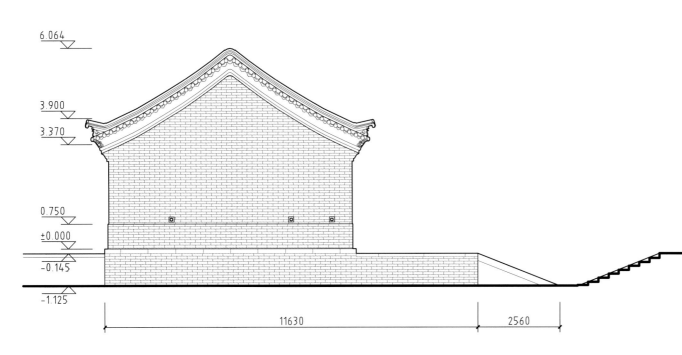

6.064

3.900
3.370

0.750
±0.000
-0.145

-1.125

11630   2560

嵩阳书院先圣殿侧立面图
Side elevation of Xianshengdian of Songyang Academy

1275 950 950 950 950 1275

6.064 垂脊顶
5.215 脊檩上皮
4.583 上金檩下皮
3.944 下金檩下皮
3.152 抱头梁下皮
2.687 穿插枋下皮
0.750
±0.000 台明
-0.145

1275   3800   1275
6350

嵩阳书院先圣殿横剖面图
Cross-section of Xianshengdian of Songyang Academy

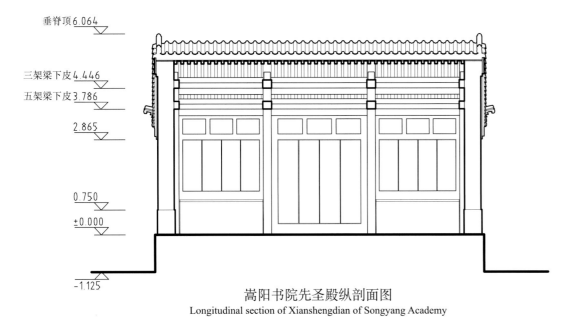

垂脊顶 6.064

三架梁下皮 4.446
五架梁下皮 3.786

2.865

0.750
±0.000

-1.125

嵩阳书院先圣殿纵剖面图
Longitudinal section of Xianshengdian of Songyang Academy

0   1   2m

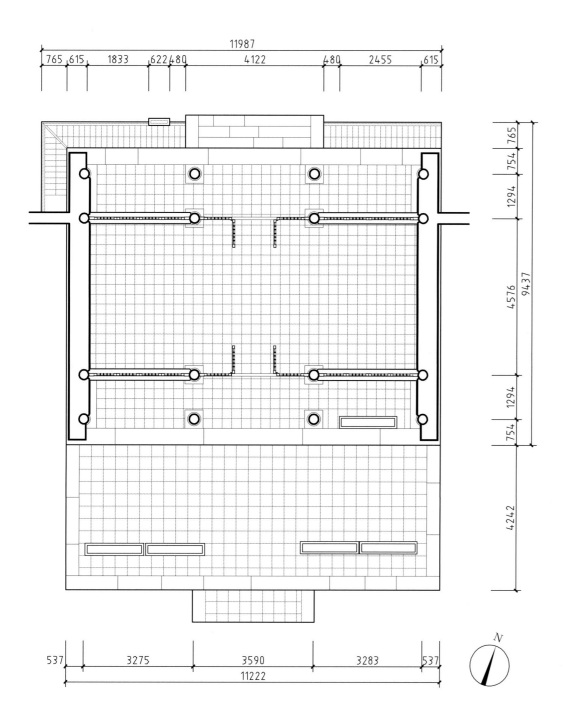

嵩阳书院讲堂平面图
Plan of *jiangtang* of Songyang Academy

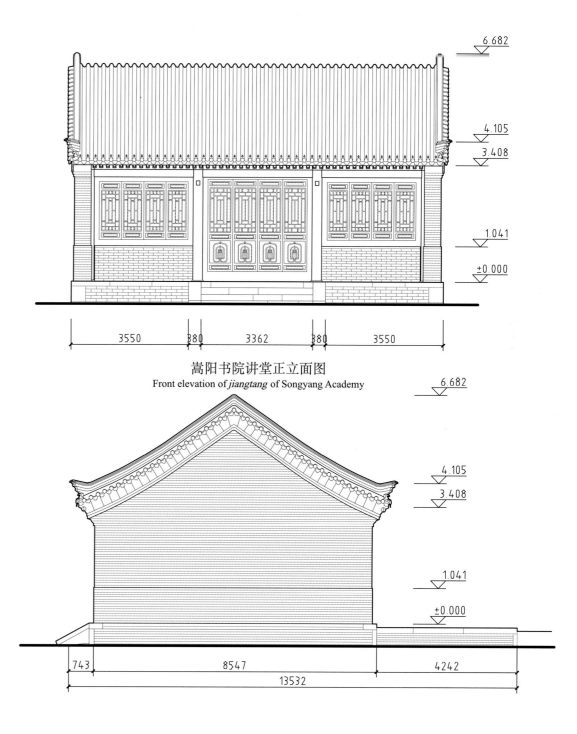

嵩阳书院讲堂正立面图
Front elevation of *jiangtang* of Songyang Academy

嵩阳书院讲堂侧立面图
Side elevation of *jiangtang* of Songyang Academy

0 1 2m

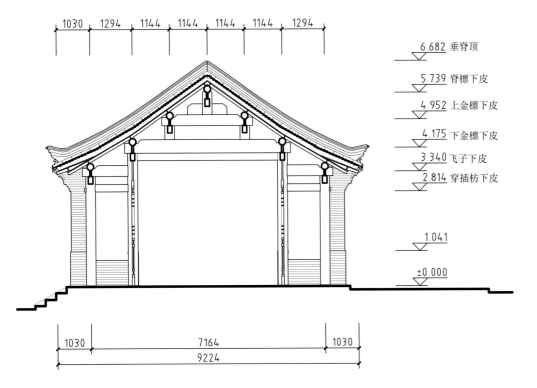

| 1030 | 1294 | 1144 | 1144 | 1144 | 1144 | 1294 |

6.682 垂脊顶

5.739 脊檩下皮

4.952 上金檩下皮

4.175 下金檩下皮

3.340 飞子下皮

2.814 穿插枋下皮

1.041

±0.000

| 1030 | 7164 | 1030 |

9224

嵩阳书院讲堂横剖面图
Cross-section of *jiangtang* of Songyang Academy

6.682 垂脊顶

4.768 三架梁下皮

3.949 五架梁下皮

±0.000 穿插枋下皮

±0.000 台明

-0.652

嵩阳书院讲堂纵剖面图
Longitudinal section of *jiangtang* of Songyang Academy

0 1 2m

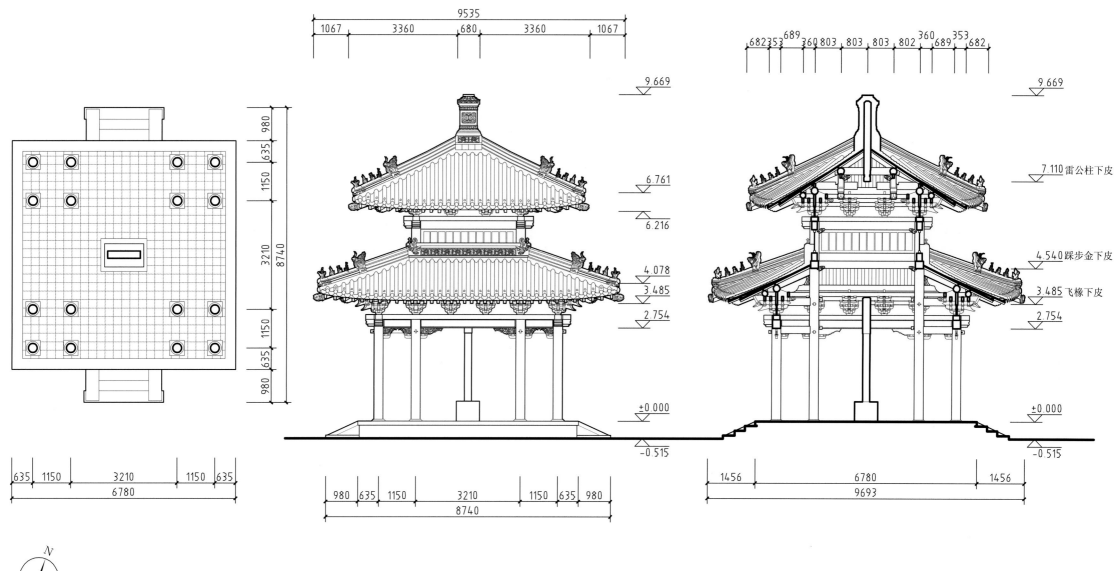

9535
1067 3360 680 3360 1067

9.669
6.761
6.216
4.078
3.485
2.754
±0.000
-0.515

682 353 689 360 803 803 803 802 360 689 353 682

9.669
7.110雷公柱下皮
4.540踩步金下皮
3.485飞椽下皮
2.754
±0.000
-0.515

980 635 1150 3210 1150 635 980
8740

1456 6780 1456
9693

635 1150 3210 1150 635
6780

980 635 1150 3210 1150 635 980
8740

N
0 1 2m

嵩阳书院碑亭平面图
Plan of Beiting of Songyang Academy

嵩阳书院碑亭立面图
Elevation of Beiting of Songyang Academy

嵩阳书院碑亭剖面图
Section of Beiting of Songyang Academy

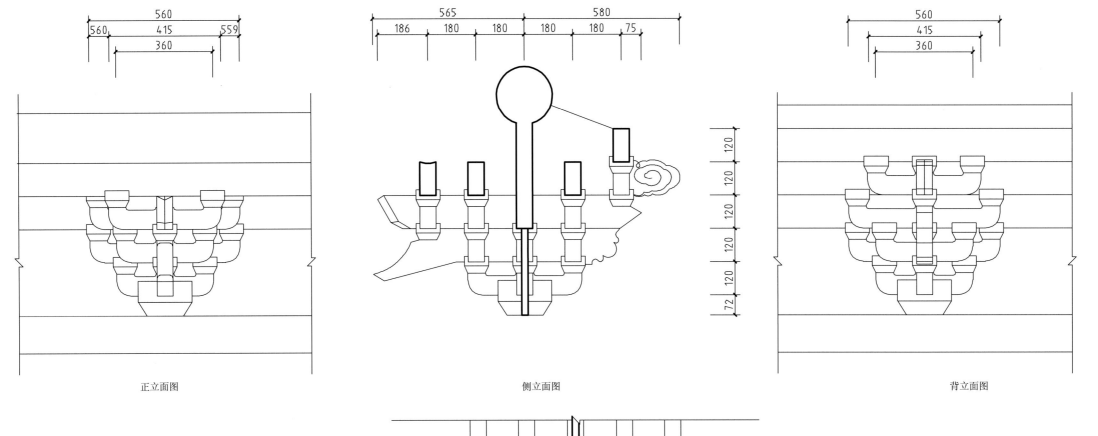

560
560 | 415 | 559
360

正立面图

565 | 580
186 | 180 | 180 | 180 | 180 | 75

120 120 120 120 120 72

侧立面图

560
415
360

背立面图

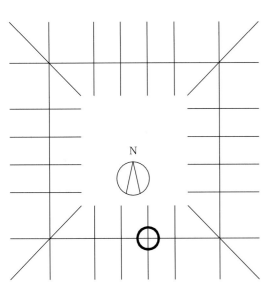

N

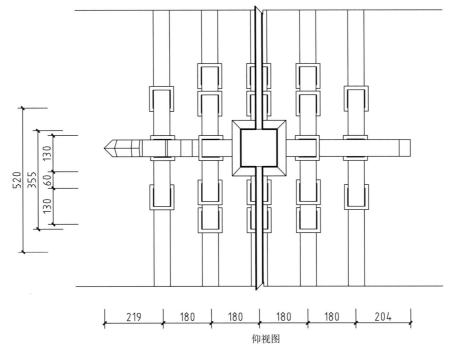

520 355 130 60 130 130

219 | 180 | 180 | 180 | 180 | 204

仰视图

嵩阳书院碑亭斗栱大样图
Bracket set of Beiting of Songyang Academy

0   0.5   1m

宝顶

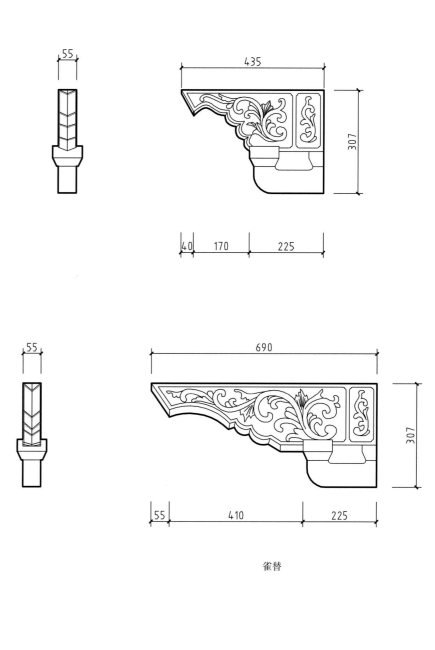

雀替

嵩阳书院碑亭构件大样图（一）
Structural components of Beiting of Songyang Academy I

0 0.1 0.2m

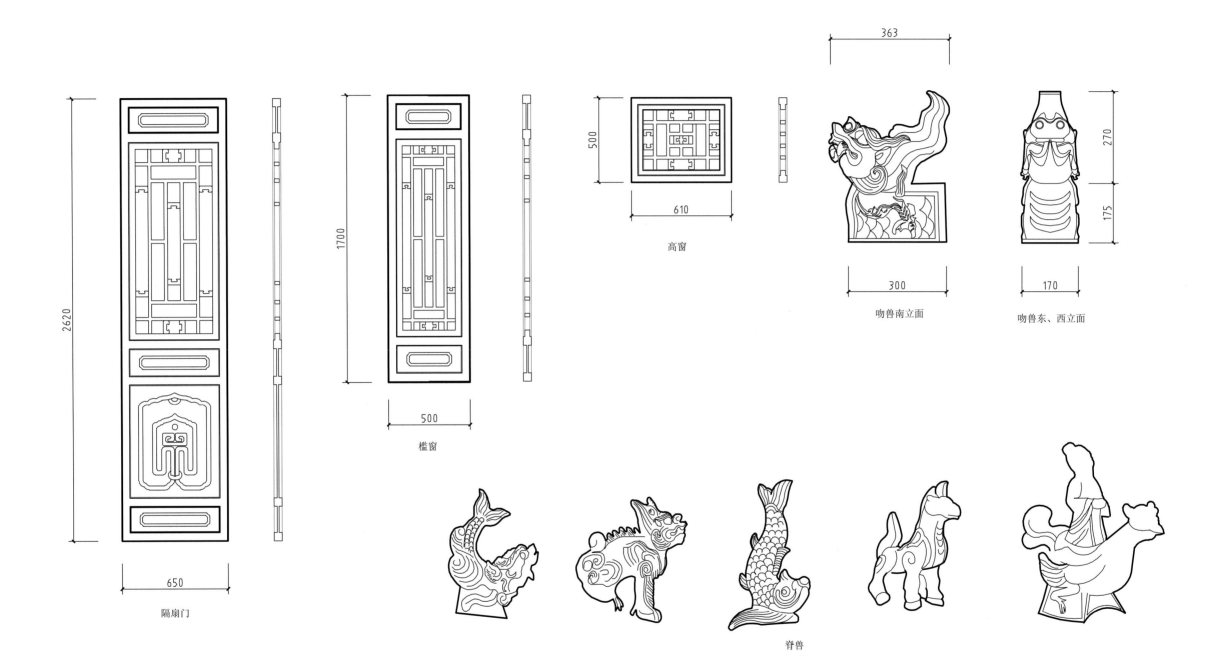

2620

650

隔扇门

1700

500

槛窗

500

610

高窗

363

300

吻兽南立面

270

175

170

吻兽东、西立面

脊兽

嵩阳书院碑亭构件大样图（二）
Structural components of Beiting of Songyang Academy II

0   0.5   1m

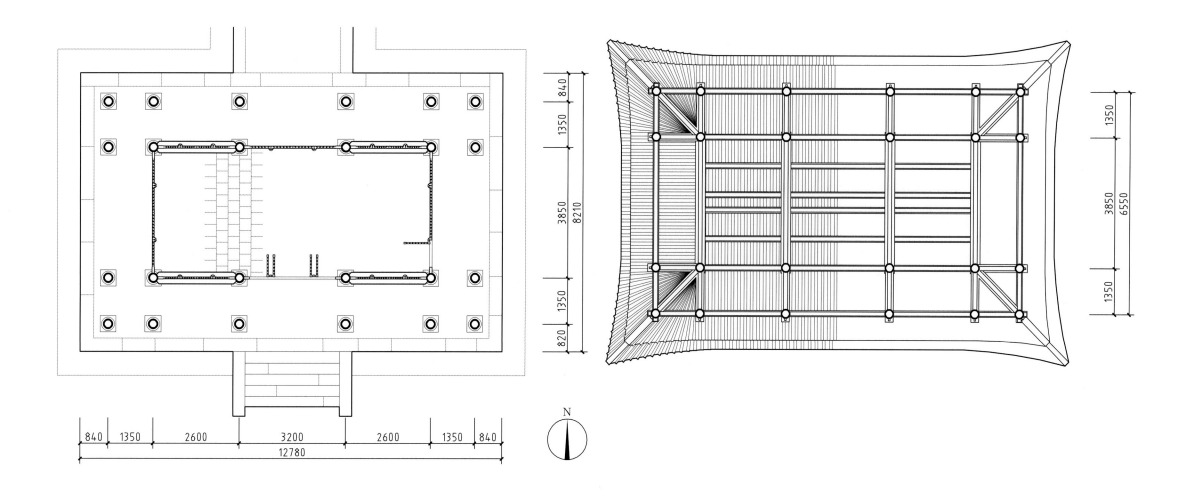

嵩阳书院道统祠首层平面图
Plan of first floor of Daotongci, Songyang Academy

嵩阳书院道统祠梁架仰视平面图
Plan of framework of Songyang Academy Daotongci as seen from below

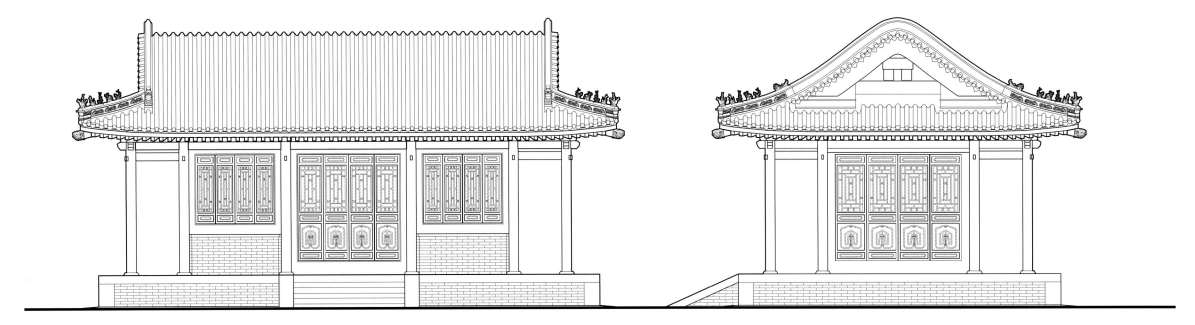

嵩阳书院道统祠正立面图
Front elevation of Daotongci of Songyang Academy

嵩阳书院道统祠侧立面图
Side elevation of Daotongci of Songyang Academy

0 1 2m

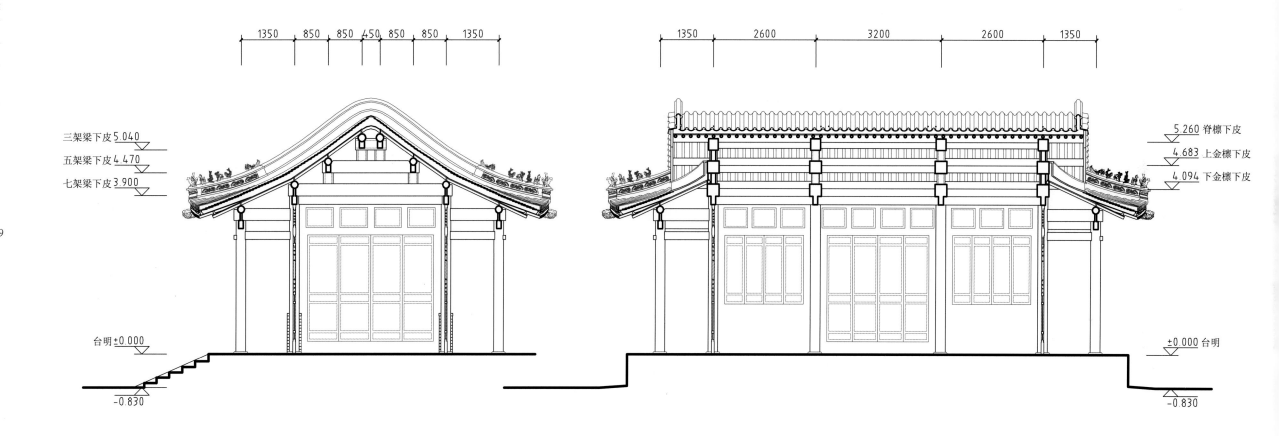

三架梁下皮5.040

五架梁下皮4.470

七架梁下皮3.900

台明±0.000

-0.830

1350 850 850 450 850 850 1350

1350 2600 3200 2600 1350

5.260 脊檩下皮

4.683 上金檩下皮

4.094 下金檩下皮

±0.000 台明

-0.830

嵩阳书院道统祠横剖面图
Cross-section of Daotongci of Songyang Academy

嵩阳书院道统祠纵剖面图
Longitudinal section of Daotongci of Songyang Academy

0   1   2m

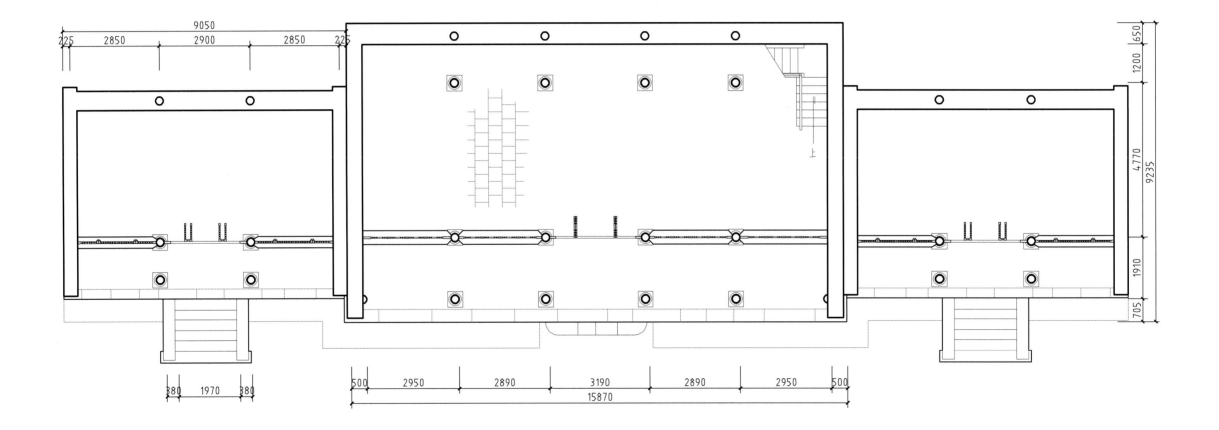

嵩阳书院藏书楼首层平面图
Plan of first floor of Cangshulou, Songyang Academy

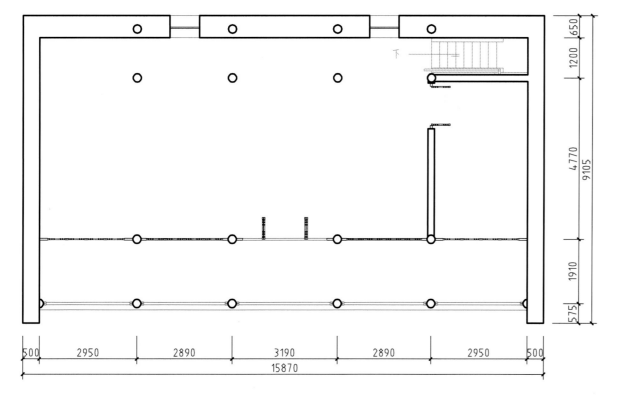

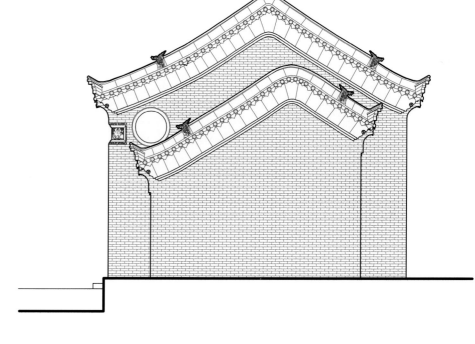

嵩阳书院藏书楼二层平面图
Plan of second floor of Cangshulou, Songyang Academy

嵩阳书院藏书楼侧立面图
Side elevation of Cangshulou, Songyang Academy

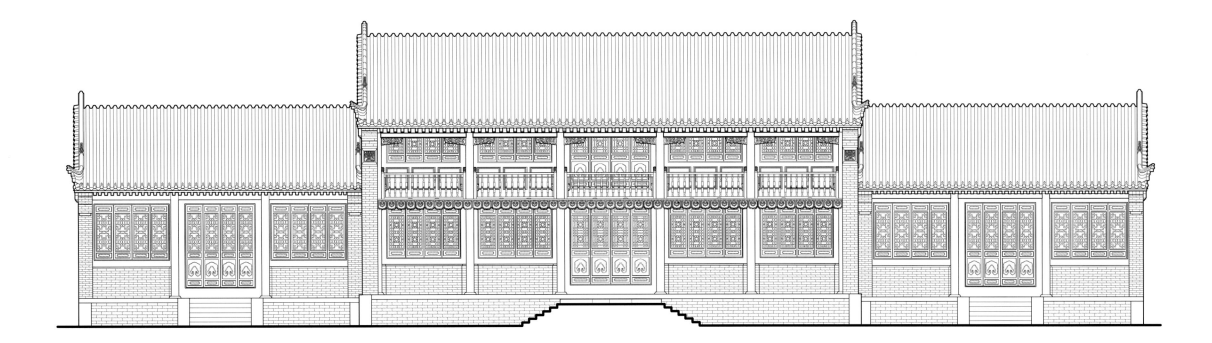

嵩阳书院藏书楼正立面图
Front elevation of Cangshulou of Songyang Academy

0 1 2m

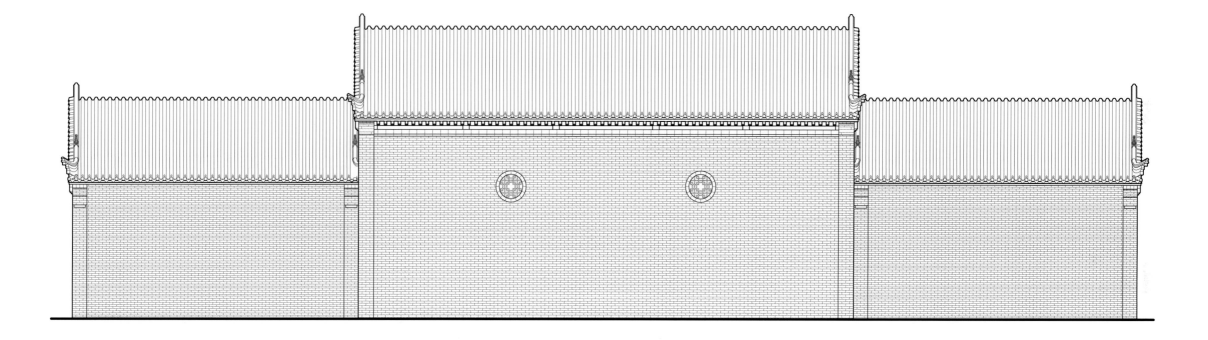

0 1 2m

嵩阳书院藏书楼背立面图
Rear elevation of Cangshulou of Songyang Academy

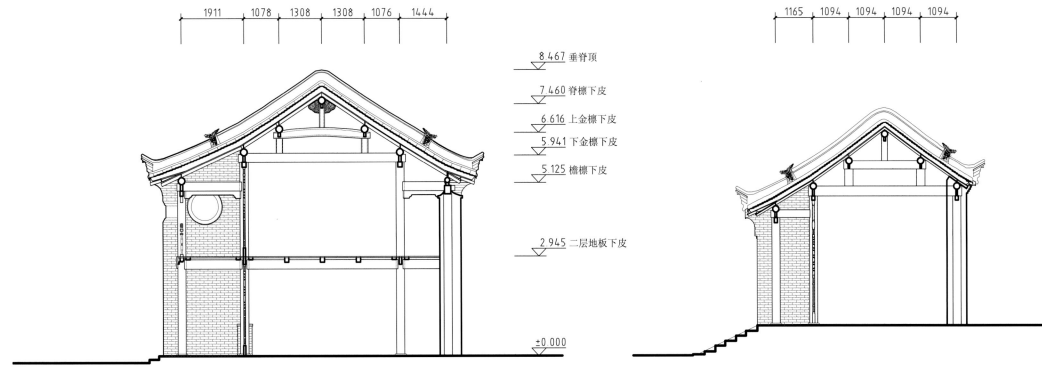

1911 1078 1308 1308 1076 1444

8.467 垂脊顶

7.460 脊檩下皮

6.616 上金檩下皮

5.941 下金檩下皮

5.125 檐檩下皮

2.945 二层地板下皮

±0.000

1165 1094 1094 1094 1094

6.436 垂脊顶

5.567 脊檩下皮

4.470 上金檩下皮

4.020 下金檩下皮

3.347 檐檩下皮

±0.000

0 1 2m

嵩阳书院藏书楼正殿横剖面图
Cross-section of *zhengdian* of Cangshulou, Songyang Academy

嵩阳书院藏书楼配殿横剖面图
Cross-section of *peidian* of Cangshulou, Songyang Academy

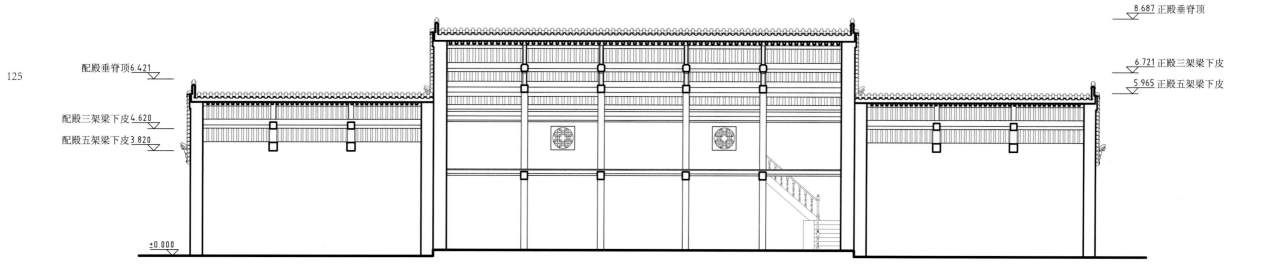

8.687 正殿垂脊顶

配殿垂脊顶6.421

6.721 正殿三架梁下皮

5.965 正殿五架梁下皮

配殿三架梁下皮4.620

配殿五架梁下皮3.820

±0.000

0　1　2m

嵩阳书院藏书楼纵剖面图
Longitudinal section of Cangshulou of Songyang Academy

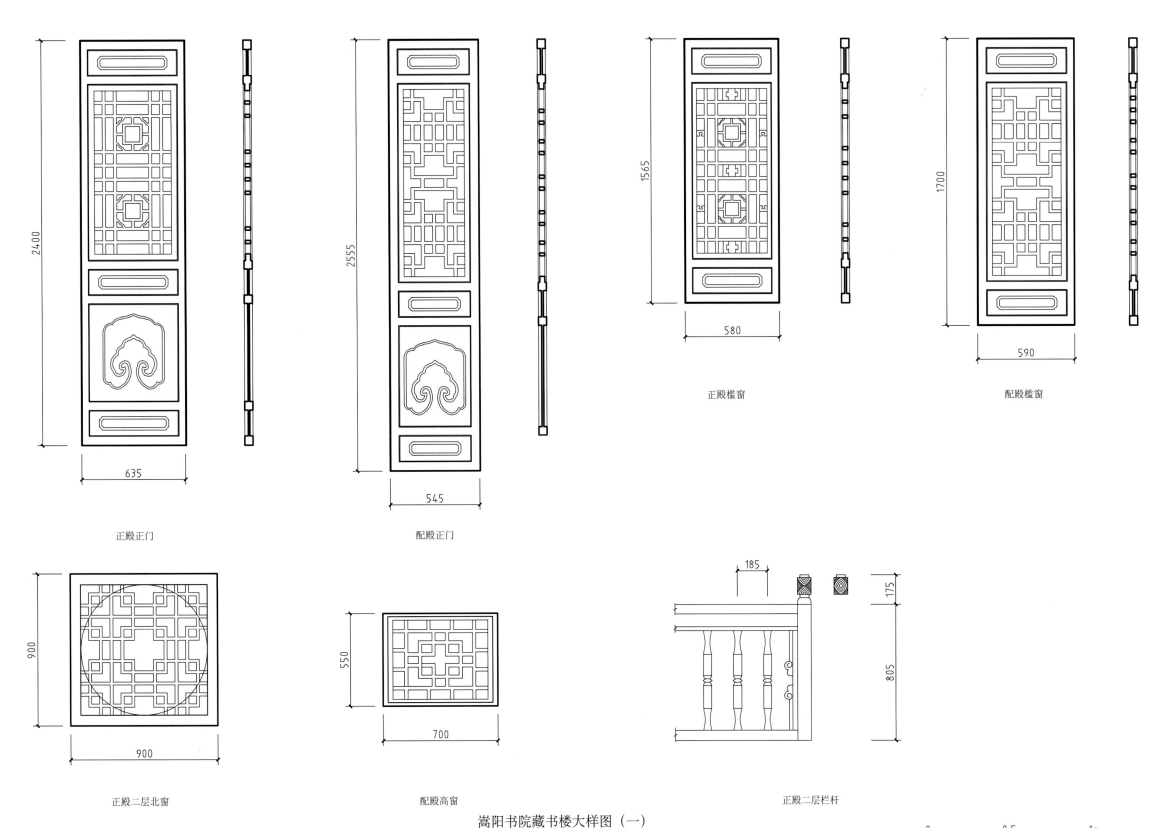

正殿正门

配殿正门

正殿槛窗

配殿槛窗

正殿二层北窗

配殿高窗

正殿二层栏杆

嵩阳书院藏书楼大样图（一）
Diagram of Cangshulou of Songyang Academy Ⅰ

0   0.5   1m

127

59

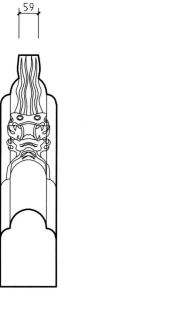

吻兽南立面

443

177

343

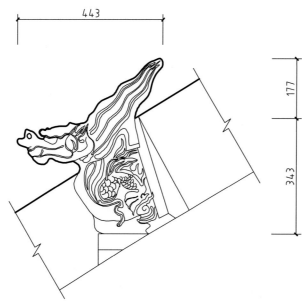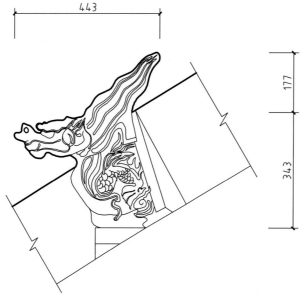

吻兽东、西立面

642

600

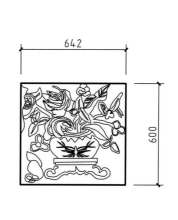

正殿山墙砖雕

1100

700

1260

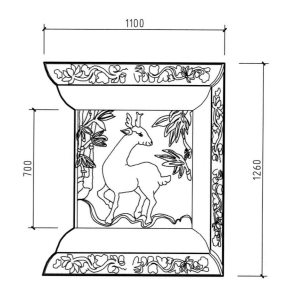

正殿山墙东、西侧砖雕

754

640

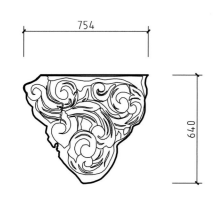

正殿梁木雕

600

590

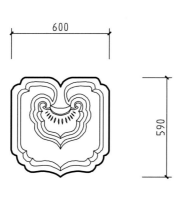

正殿挂檐板

1400

455

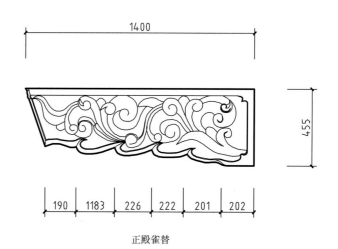

190 1183 226 222 201 202

正殿雀替

嵩阳书院藏书楼大样图（二）
Diagram of Cangshulou of Songyang Academy Ⅱ

0   0.5   1m

中岳庙
Zhongyue Temple

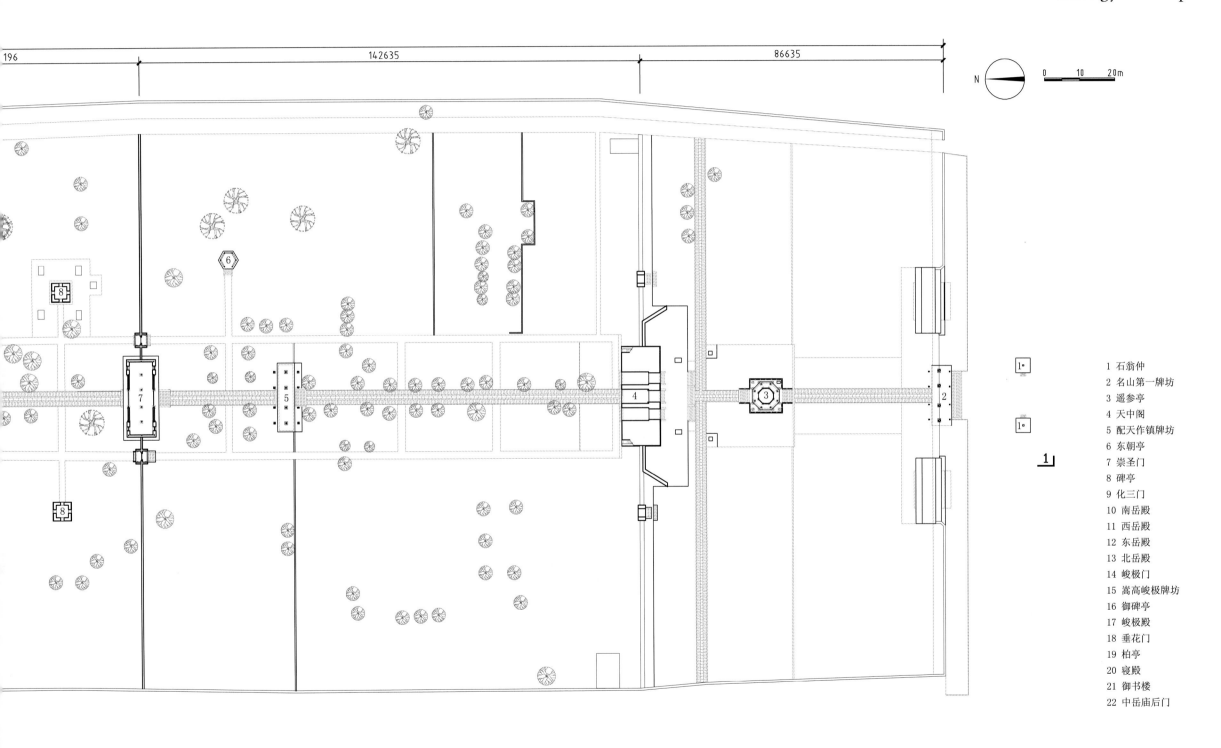

196   142635   86635

N   0   10   20m

1 石翁仲
2 名山第一牌坊
3 遥参亭
4 天中阁
5 配天作镇牌坊
6 东朝亭
7 崇圣门
8 碑亭
9 化三门
10 南岳殿
11 西岳殿
12 东岳殿
13 北岳殿
14 峻极门
15 嵩高峻极牌坊
16 御碑亭
17 峻极殿
18 垂花门
19 柏亭
20 寝殿
21 御书楼
22 中岳庙后门

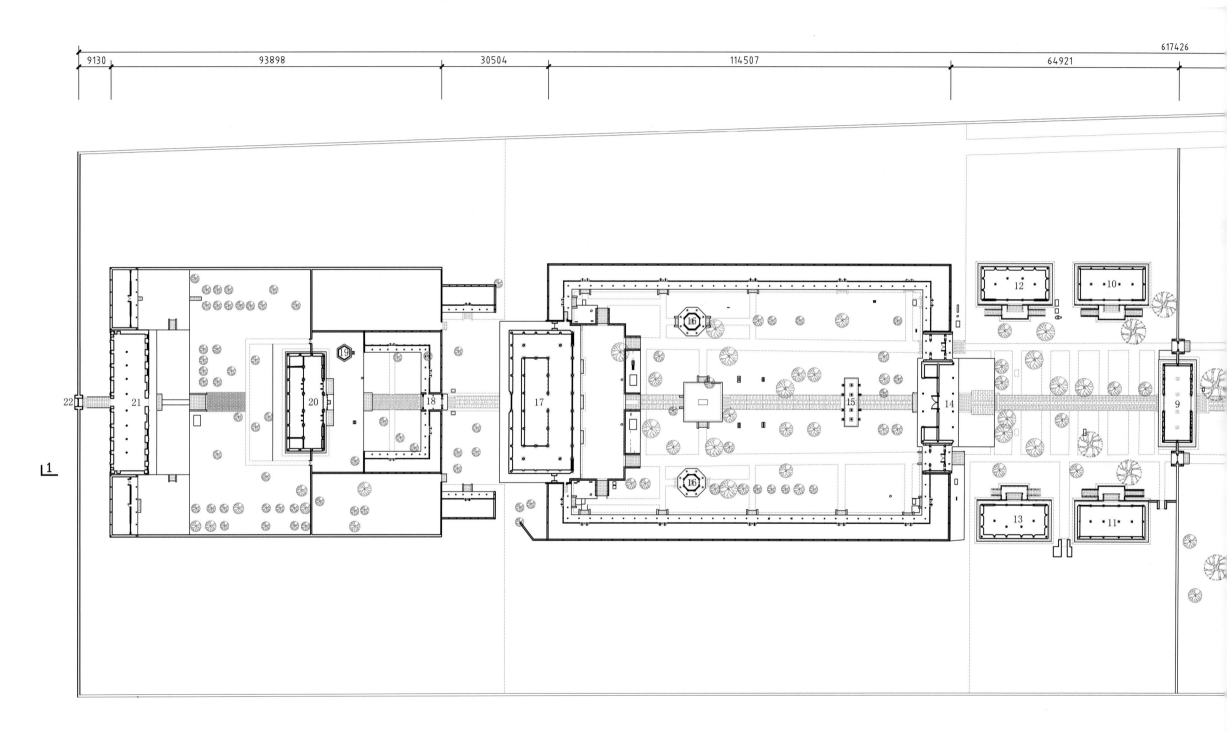

9130   93898   30504   114507   64921

617426

中岳庙总平面图
Site plan of Zhongyue Temple

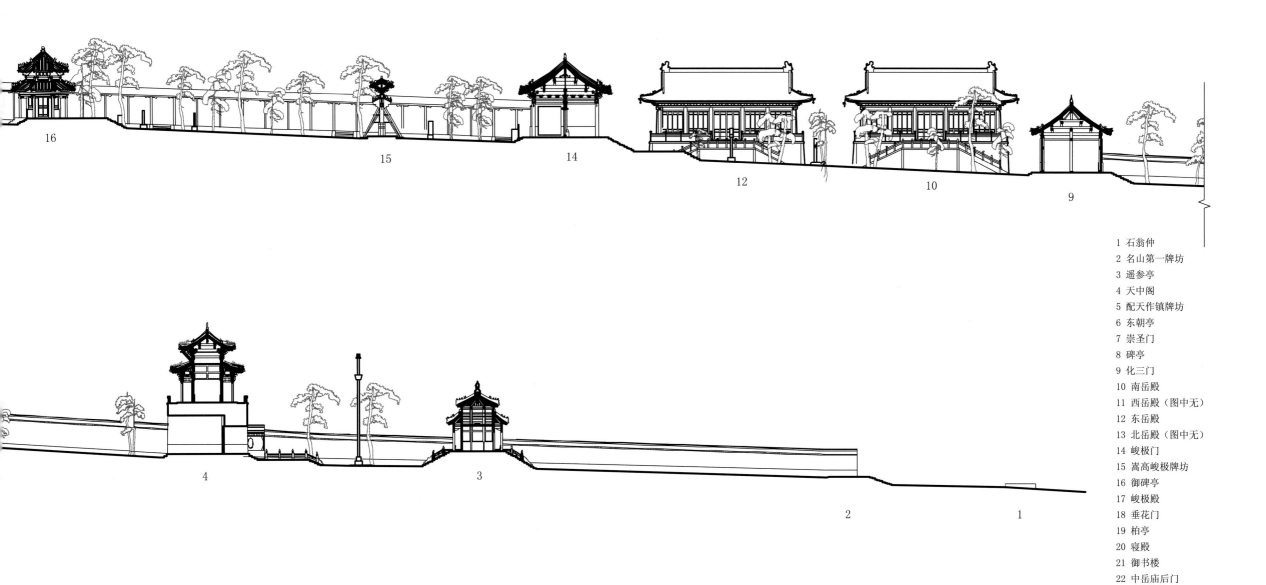

1 石翁仲
2 名山第一牌坊
3 遥参亭
4 天中阁
5 配天作镇牌坊
6 东朝亭
7 崇圣门
8 碑亭
9 化三门
10 南岳殿
11 西岳殿（图中无）
12 东岳殿
13 北岳殿（图中无）
14 峻极门
15 嵩高峻极牌坊
16 御碑亭
17 峻极殿
18 垂花门
19 柏亭
20 寝殿
21 御书楼
22 中岳庙后门

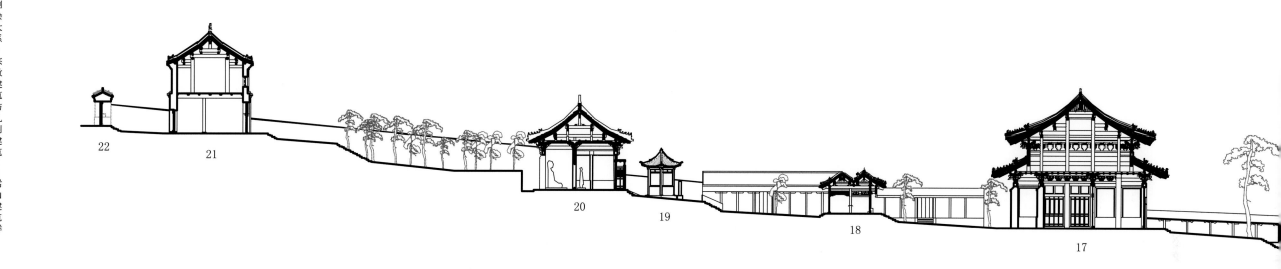

22   21   20   19   18   17

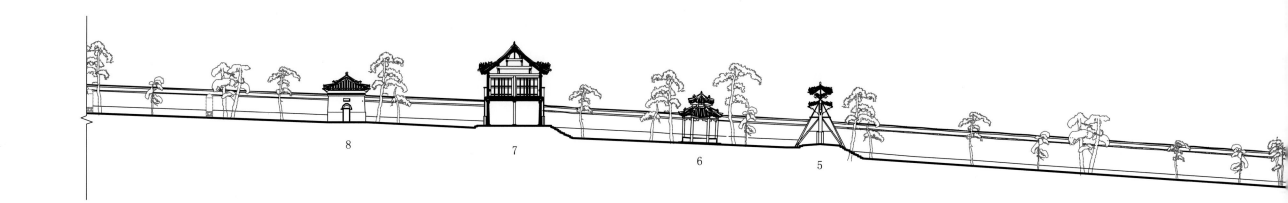

8   7   6   5

中岳庙 1-1 总剖面图
Section 1-1 of Zhongyue Temple

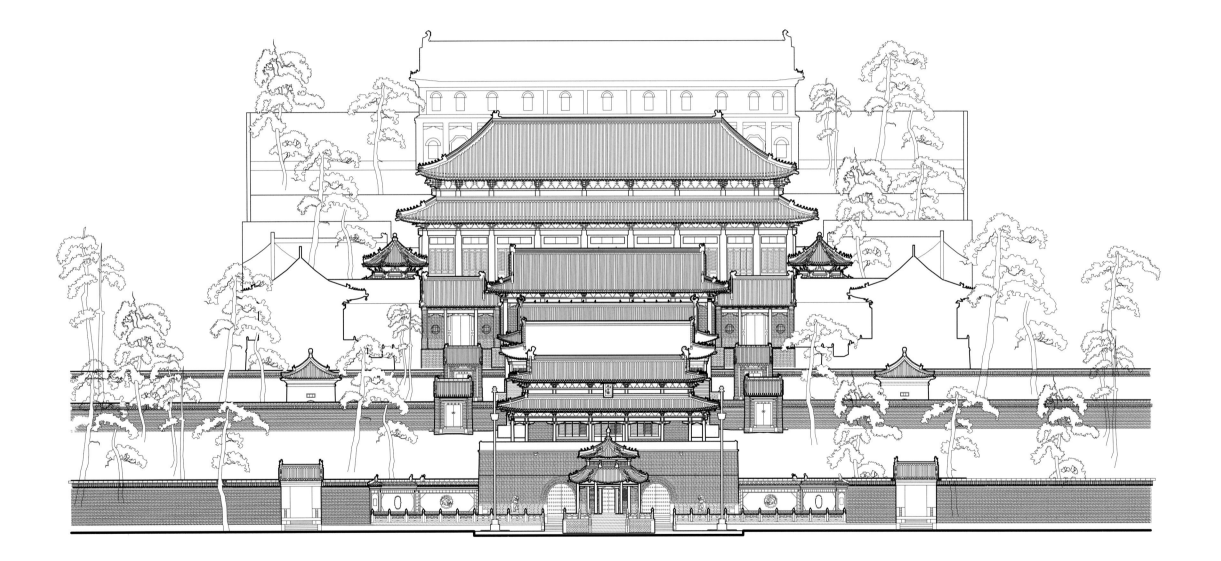

中岳庙总正立面图
Front elevation of Zhongyue Temple

0　1　　　　5m

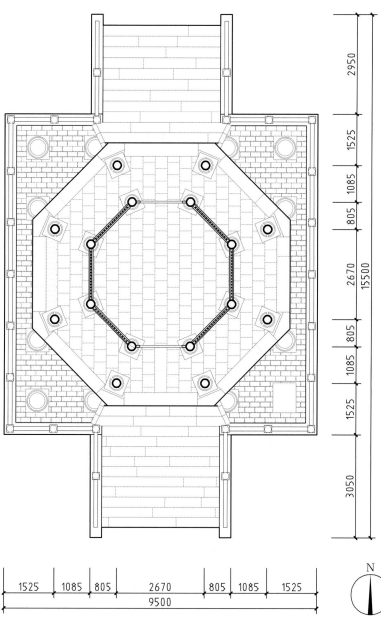

中岳庙遥参亭首层平面图
Plan of first floor of Yaocanting, Zhongyue Temple

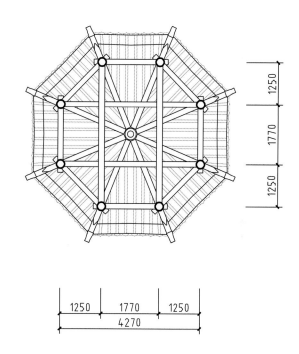

中岳庙遥参亭梁架仰视平面图
Plan of framework of Zhongyue Temple Yaocanting as seen from below

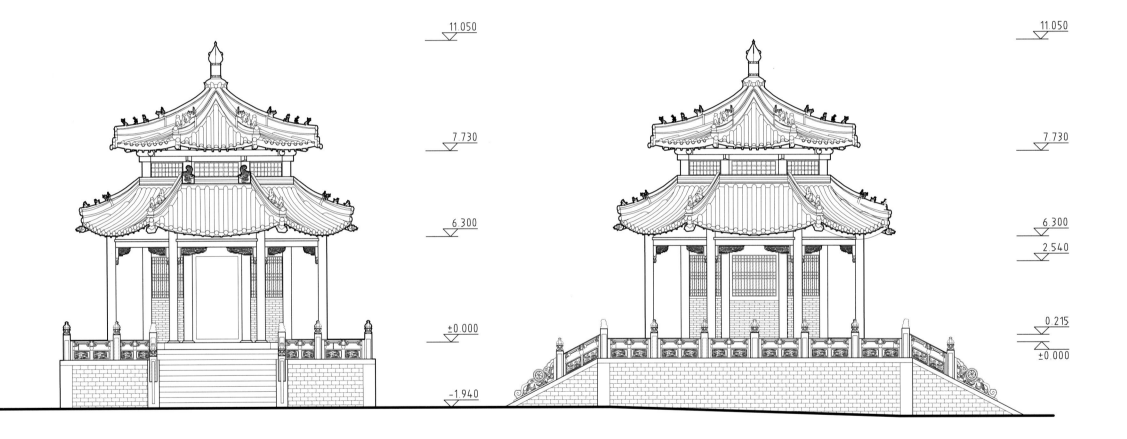

中岳庙遥参亭正立面图
Front elevation of Yaocanting of Zhongyue Temple

中岳庙遥参亭侧立面图
Side elevation of Yaocanting of Zhongyue Temple

0　1　2m

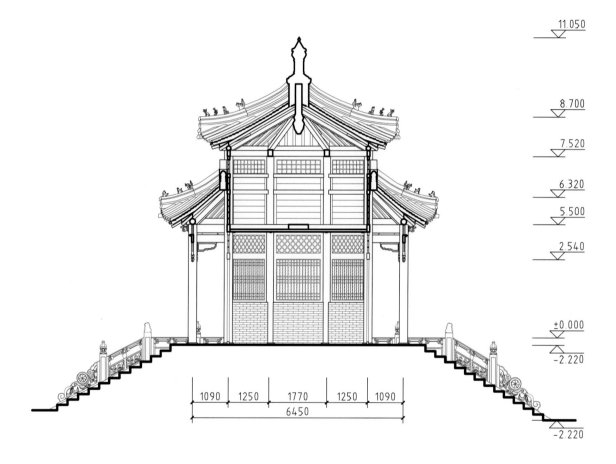

11.050

8.700

7.520

6.320

5.500

2.540

±0.000

-2.220

1090　1250　1770　1250　1090

6450

-2.220

中岳庙遥参亭横剖面图

Cross-section of Yaocanting of Zhongyue Temple

0　1　2m

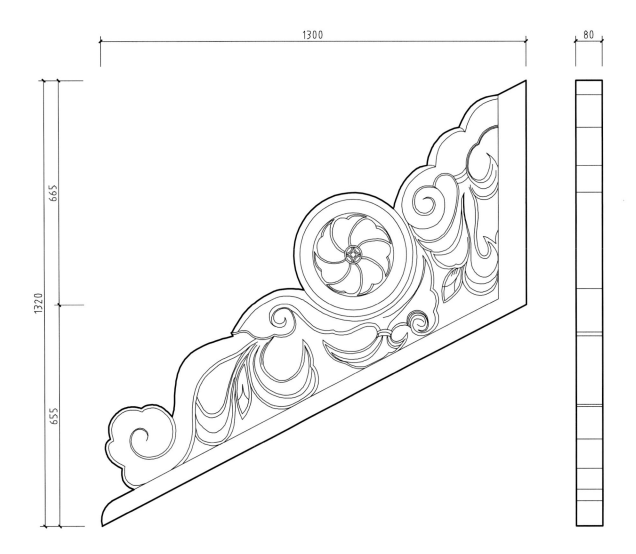

中岳庙遥参亭雀替大样图
Queti of Yaocanting of Zhongyue Temple

中岳庙遥参亭抱鼓石大样图
Baogushi of Yaocanting of Zhongyue Temple

0   1   2m

围脊吻兽

上层戗脊走兽

下层戗脊走兽

中岳庙遥参亭构件大样图
Structural components of Yaocanting of Zhongyue Temple

0   0.1   0.2m

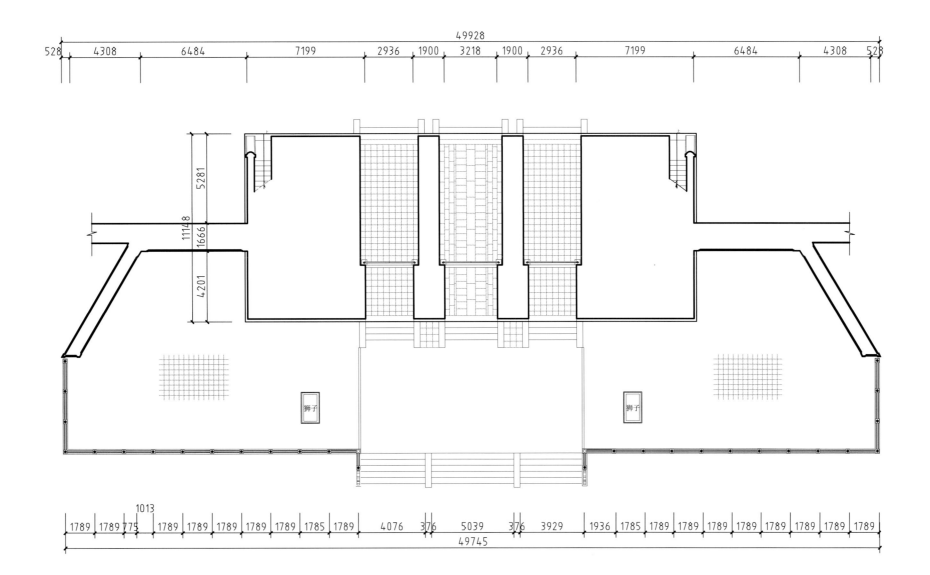

中岳庙天中阁总平面图
Site plan of Tianzhongge of Zhongyue Temple

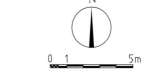

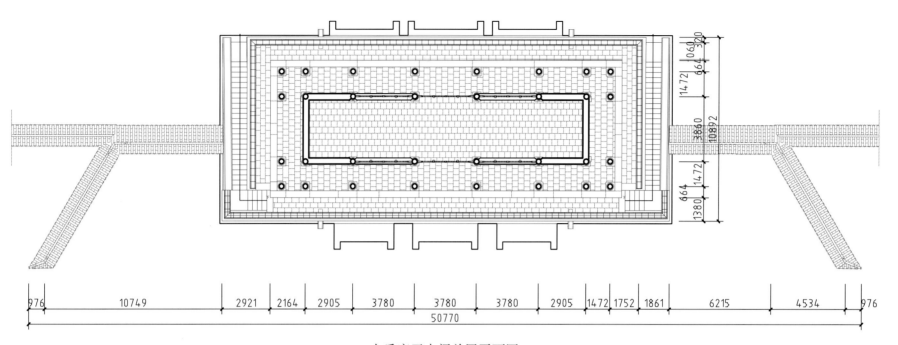

中岳庙天中阁首层平面图
Plan of first floor of Tianzhongge, Zhongyue Temple

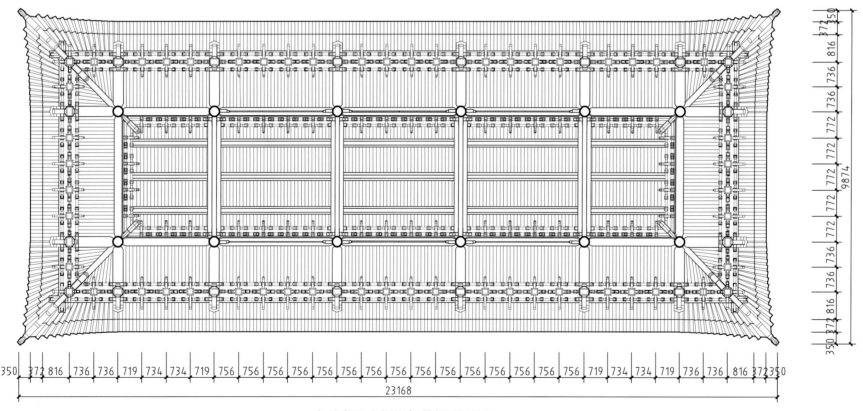

中岳庙天中阁梁架仰视平面图
Plan of framework of Zhongyue Temple Tianzhongge as seen from below

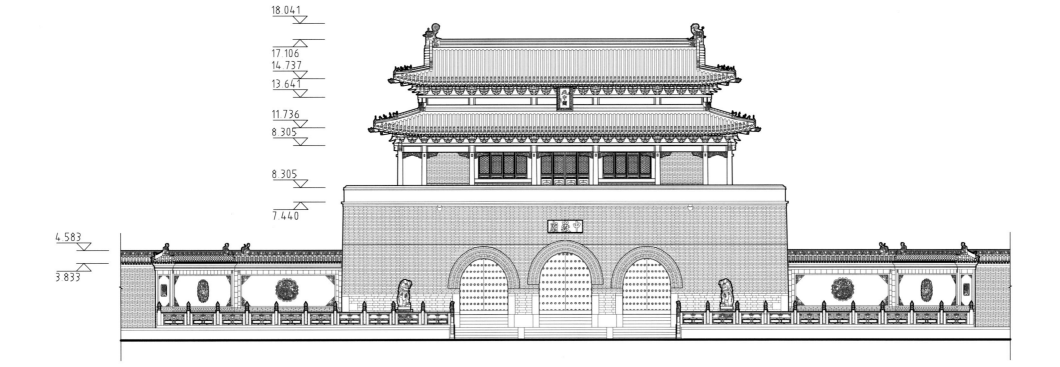

18.041

17.106
14.737
13.641

11.736
8.305

8.305

7.440

4.583

3.833

中岳庙天中阁总正立面图
Site elevation of front of Tianzhongge, Zhongyue Temple

0 1   5m

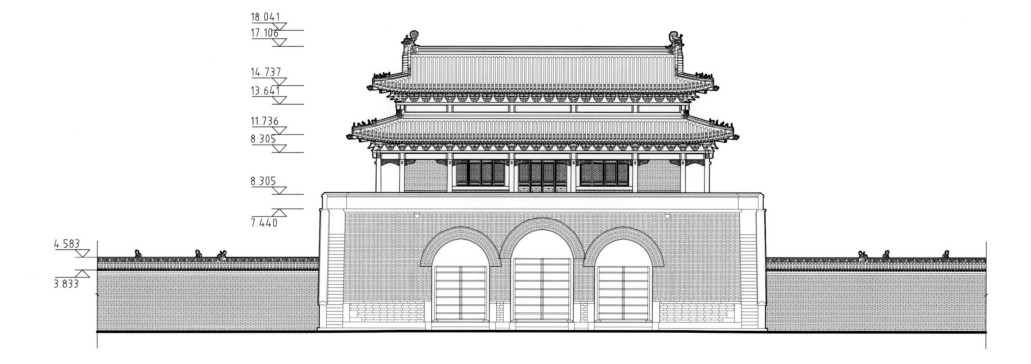

中岳庙天中阁总背立面图
Site elevation of rear of Tianzhongge, Zhongyue Temple

0 1   5m

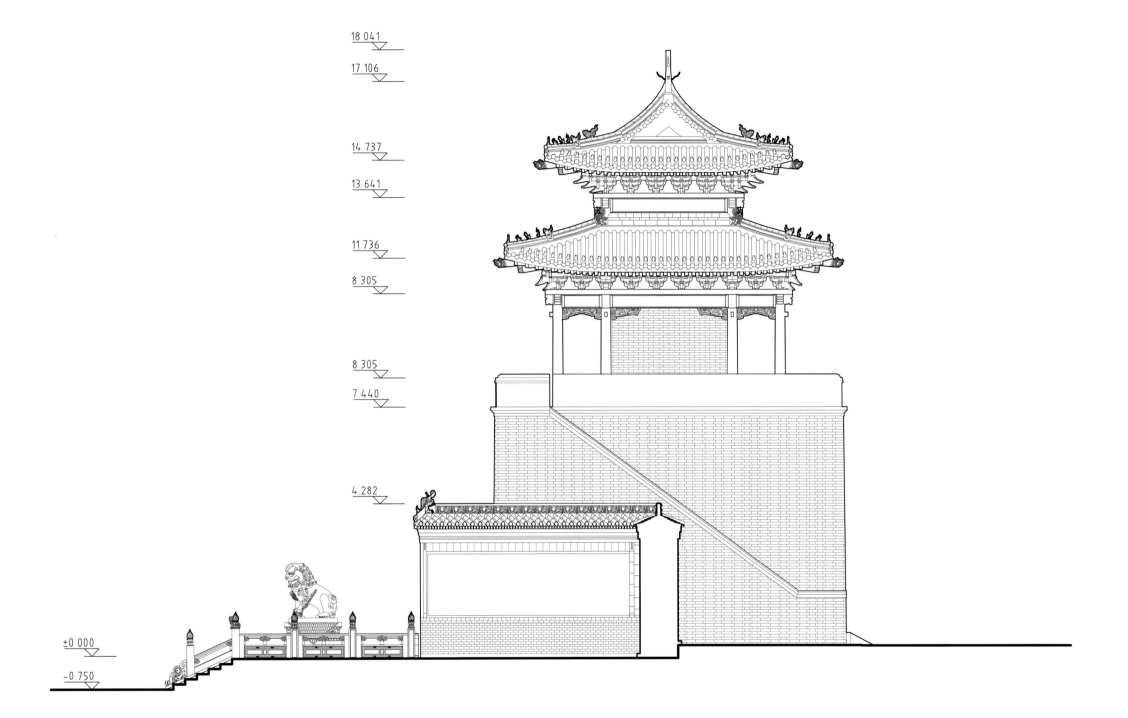

18.041

17.106

14.737

13.641

11.736

8.305

8.305

7.440

4.282

±0.000

−0.750

中岳庙天中阁总侧立面图
Site elevation of side elevation of Tianzhongge, Zhongyue Temple

0   1   2m

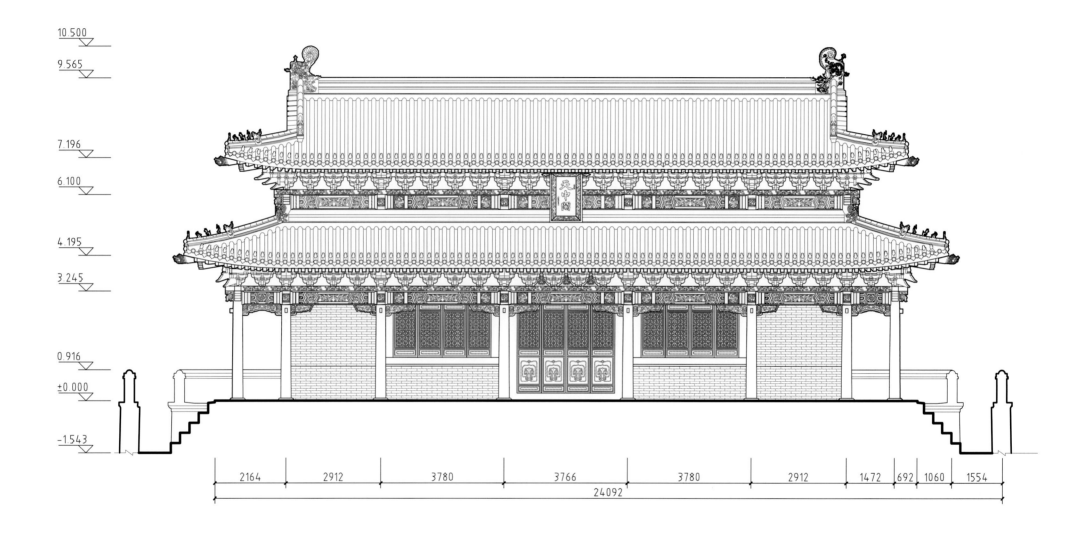

10.500

9.565

7.196

6.100

4.195

3.245

0.916

±0.000

-1.543

2164　2912　3780　3766　3780　2912　1472　692　1060　1554

24092

中岳庙天中阁正立面图
Front elevation of Tianzhongge of Zhongyue Temple

0　1　2m

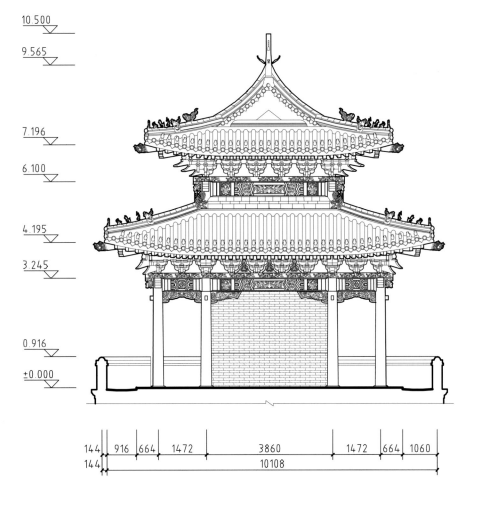

10.500

9.565

7.196

6.100

4.195

3.245

0.916

±0.000

| 144 | 916 | 664 | 1472 | 3860 | 1472 | 664 | 1060 |

144　10108

中岳庙天中阁侧立面图
Side elevation of Tianzhongge of Zhongyue Temple

0　1　2m

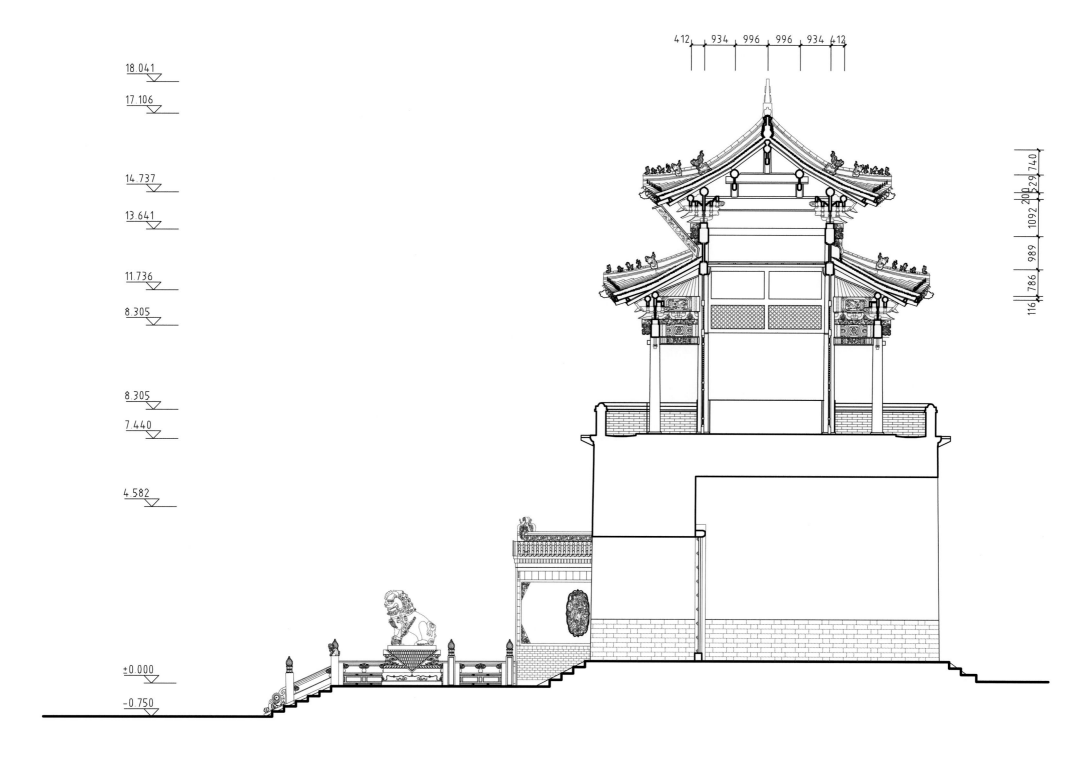

中岳庙天中阁总横剖面图
Cross-section of Tianzhongge of Zhongyue Temple

0 1 2m

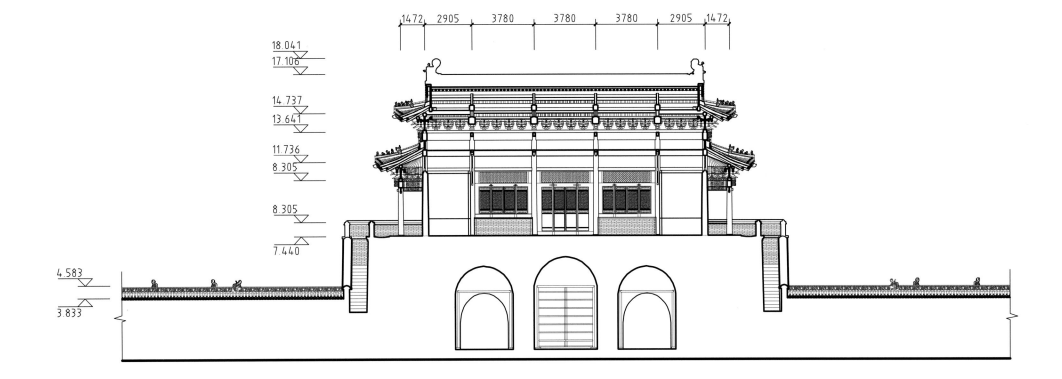

1472 2905 3780 3780 3780 2905 1472

18.041
17.106
14.737
13.641
11.736
8.305
8.305
7.440
4.583
3.833

中岳庙天中阁总纵剖面图
Longitudinal section of Tianzhongge of Zhongyue Temple

0 1 5m

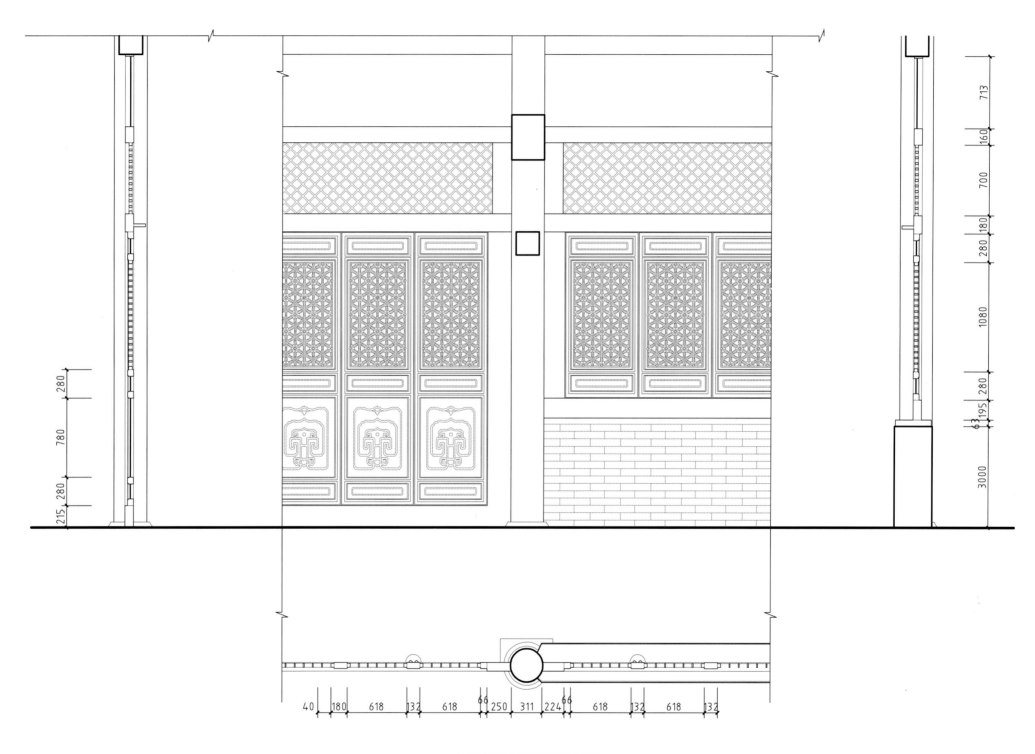

中岳庙天中阁门窗大样图
Doors and windows of Tianzhongge of Zhongyue Temple

0   0.5m

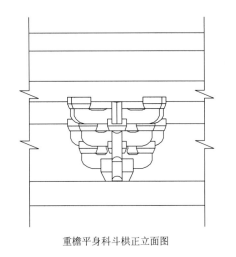

重檐平身科斗栱正立面图

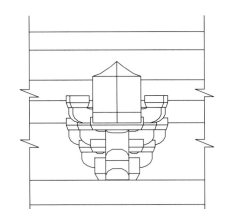

重檐柱头科斗栱正立面图

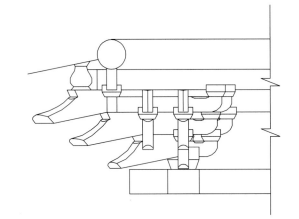

重檐柱头科斗栱立面图

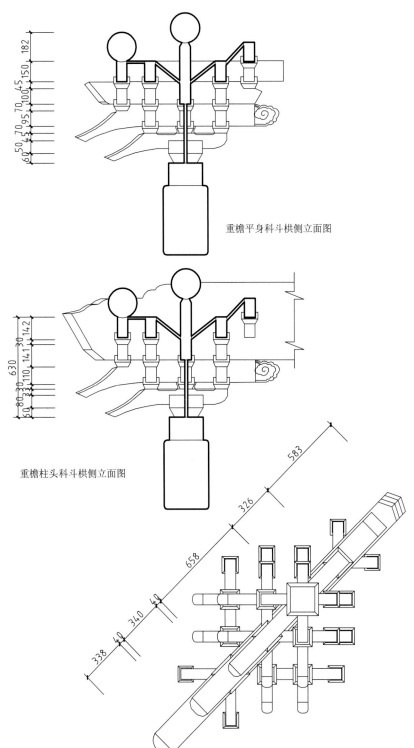

重檐平身科斗栱侧立面图

重檐柱头科斗栱侧立面图

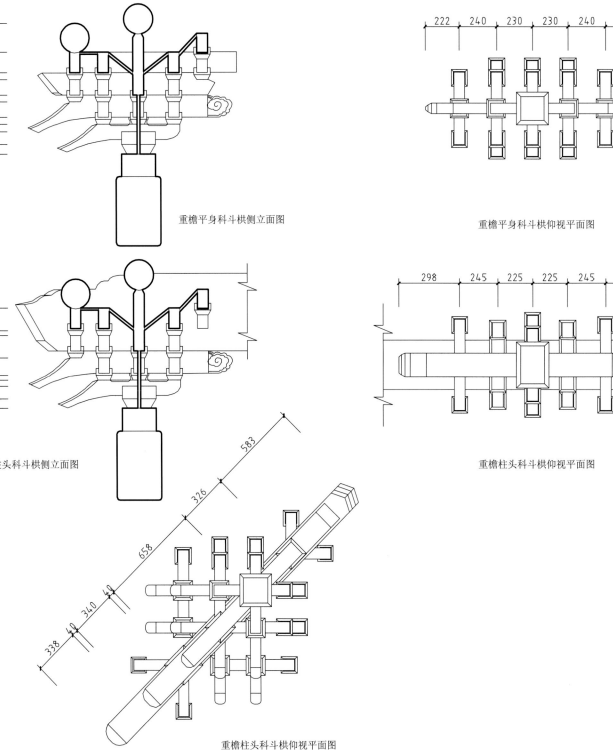

重檐柱头科斗栱仰视平面图

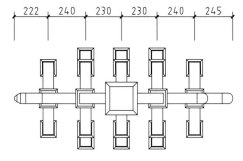

重檐平身科斗栱仰视平面图

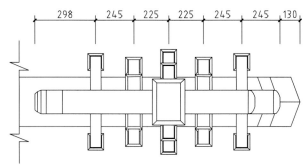

重檐柱头科斗栱仰视平面图

中岳庙天中阁斗栱大样图（一）
Bracket set of Tianzhongge of Zhongyue Temple I

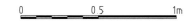

0   05   1m

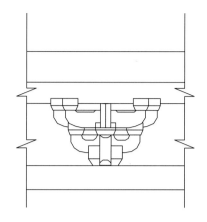

下檐平身科斗栱正立面图

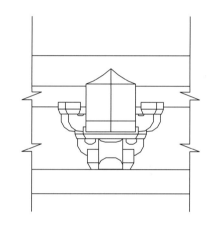

下檐柱头科斗栱正立面图

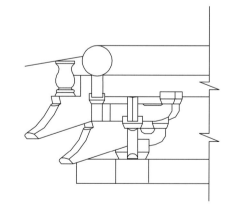

下檐柱头科斗栱立面图

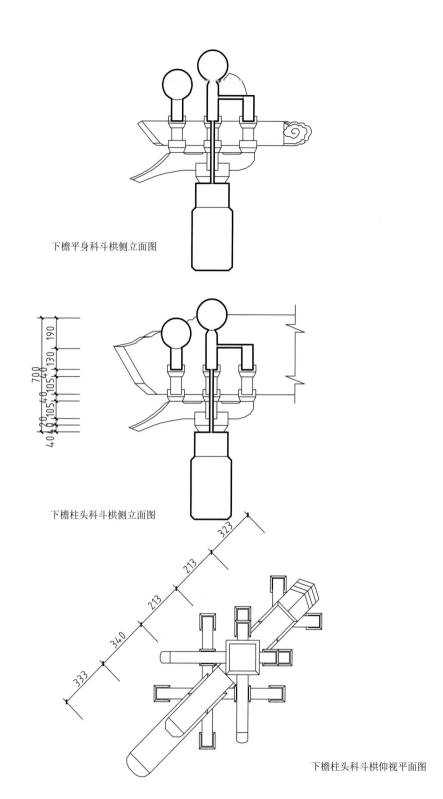

下檐平身科斗栱侧立面图

下檐柱头科斗栱侧立面图

下檐柱头科斗栱仰视平面图

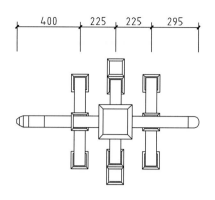

下檐平身科斗栱仰视平面图

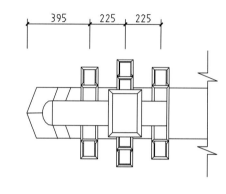

下檐柱头科斗栱仰视平面图

中岳庙天中阁斗栱大样图（二）
Bracket set of Tianzhongge of Zhongyue Temple II

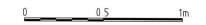

0　　　0.5　　　1m

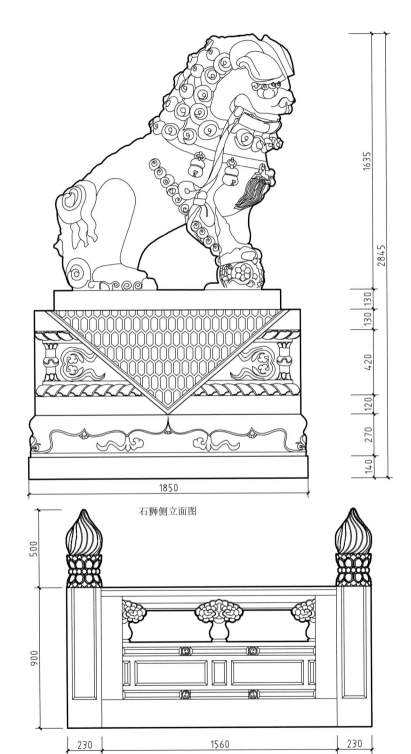

石狮侧立面图

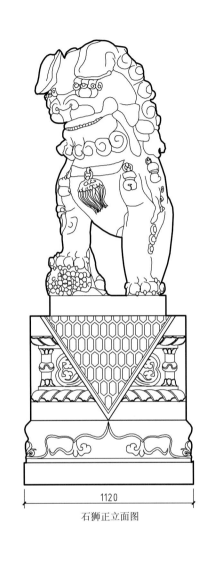

石狮正立面图

影壁图案

影壁图案

栏杆

抱鼓石

影壁图案

影壁图案

中岳庙天中阁构件大样图
Structural components of Tianzhongge of Zhongyue Temple

0   0.5   1m

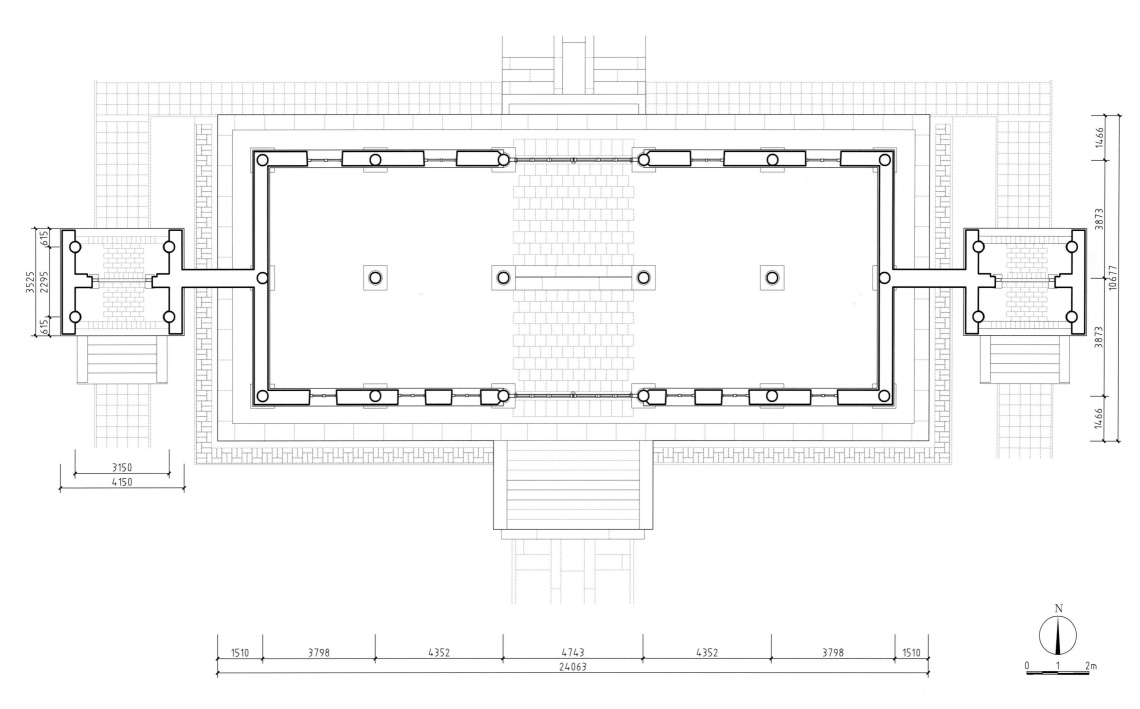

中岳庙化三门平面图
Plan of Huasanmen of Zhongyue Temple

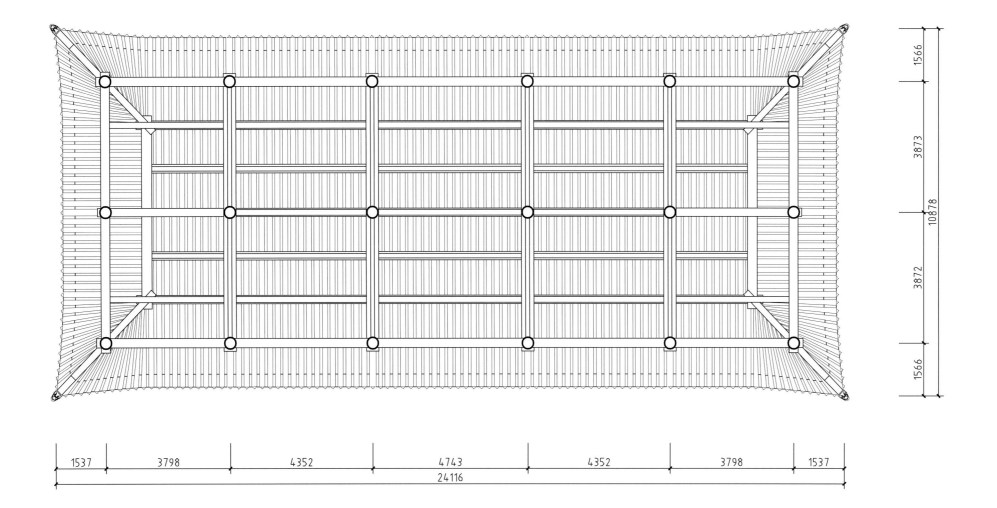

中岳庙化三门梁架仰视平面图
Plan of framework of Zhongyue Temple Huasanmen as seen from below

0   1   2m

9.130

5.900

4.740

2.780

2.320

1.350

±0.000

−0.500

−1.320

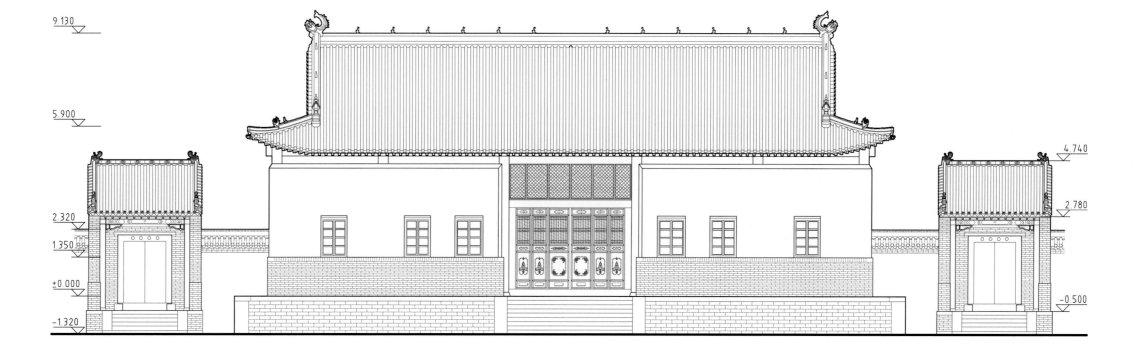

中岳庙化三门正立面图
Front elevation of Huasanmen of Zhongyue Temple

0 1 2m

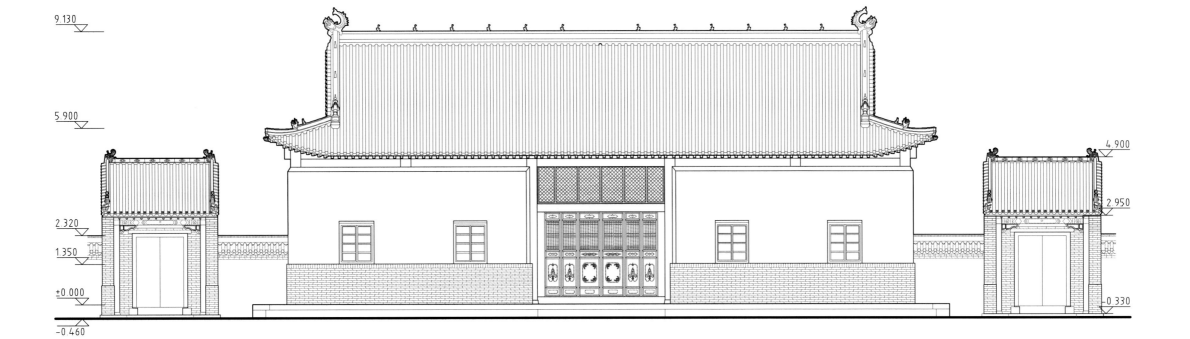

9.130

5.900

4.900

2.950

2.320

1.350

±0.000

-0.330

-0.460

中岳庙化三门背立面图
Rear elevation of Huasanmen of Zhongyue Temple

0   1   2m

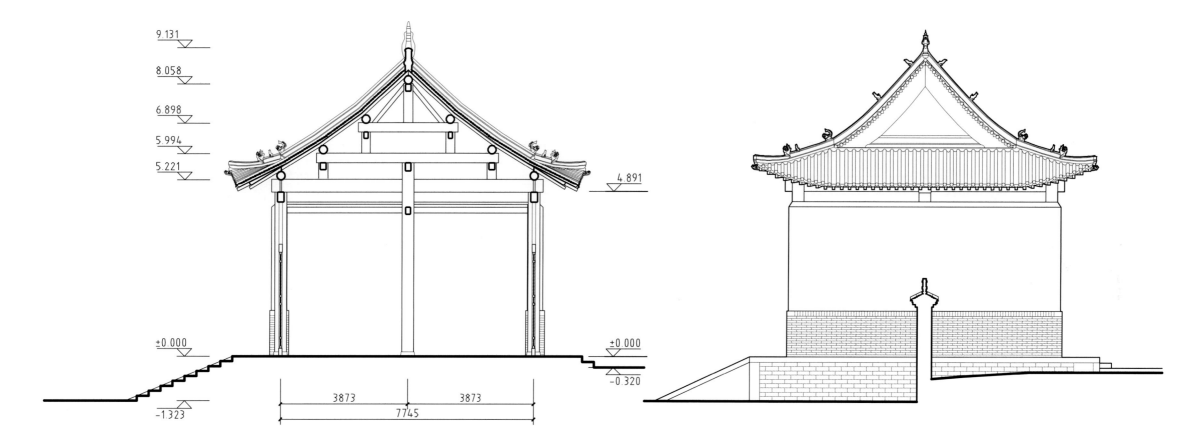

中岳庙化三门横剖面图
Cross-section of Huasanmen of Zhongyue Temple

中岳庙化三门侧立面图
Side elevation of Huasanmen of Zhongyue Temple

9.131

5.221

±0.000

-0.686

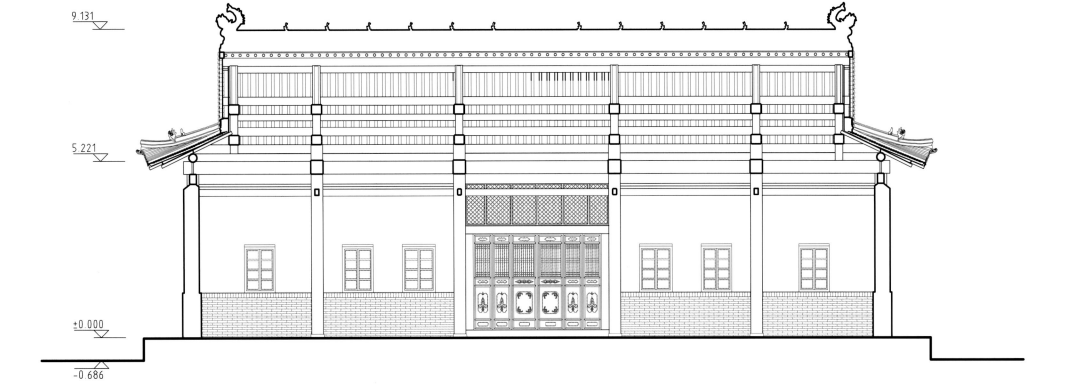

中岳庙化三门纵剖面图
Longitudinal section of Huasanmen of Zhongyue Temple

0   1   2m

小门额枋

小门正脊兽

正房脊兽

正房正脊兽

小门走兽

小门走兽

中岳庙化三门构件大样图
Structural components of Huasanmen of Zhongyue Temple

0　0.1　0.2m

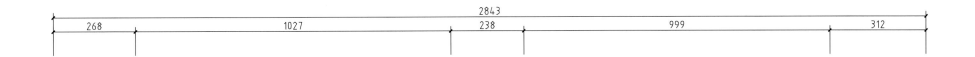

中岳庙化三门门扇大样图
Doors and windows of Huasanmen of Zhongyue Temple

0 0.1 0.2m

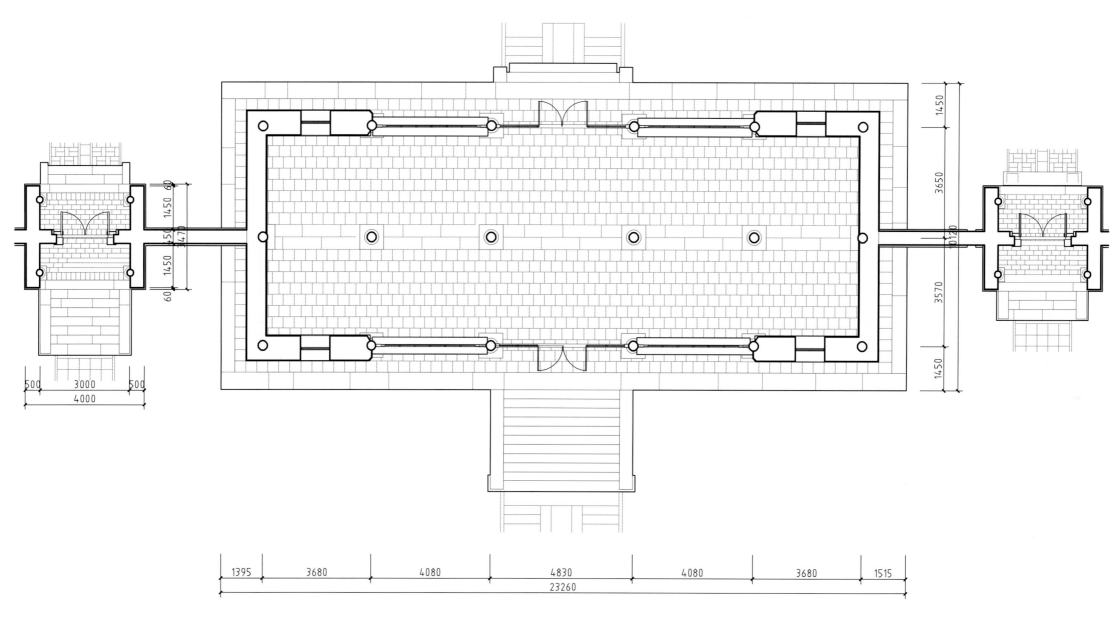

中岳庙崇圣门首层平面图
Plan of first floor of Chongshengmen of Zhongyue Temple

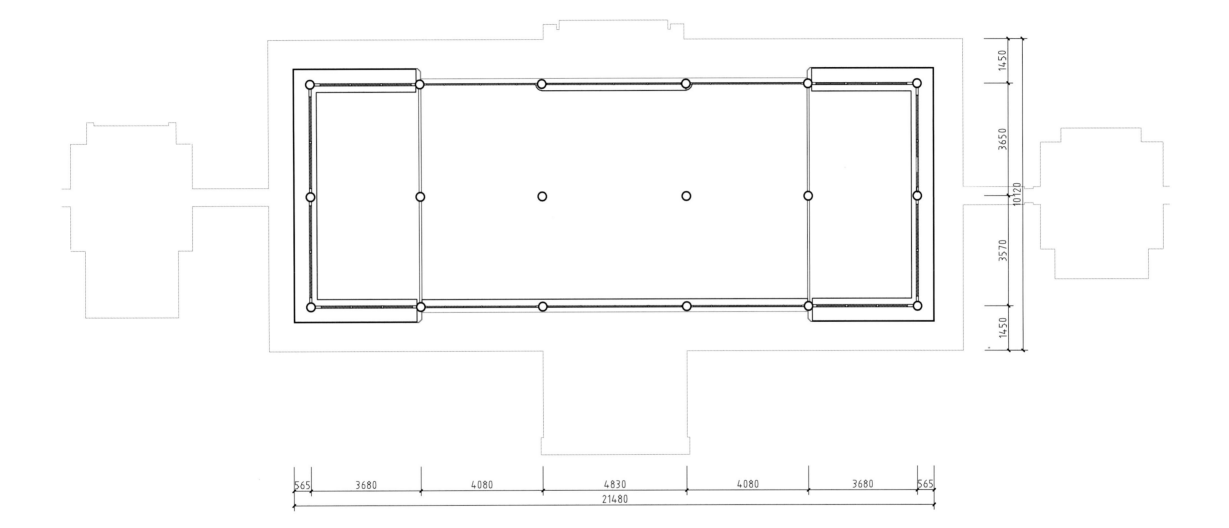

中岳庙崇圣门二层平面图
Plan of second floor of Chongshengmen of Zhongyue Temple

0   1   2m

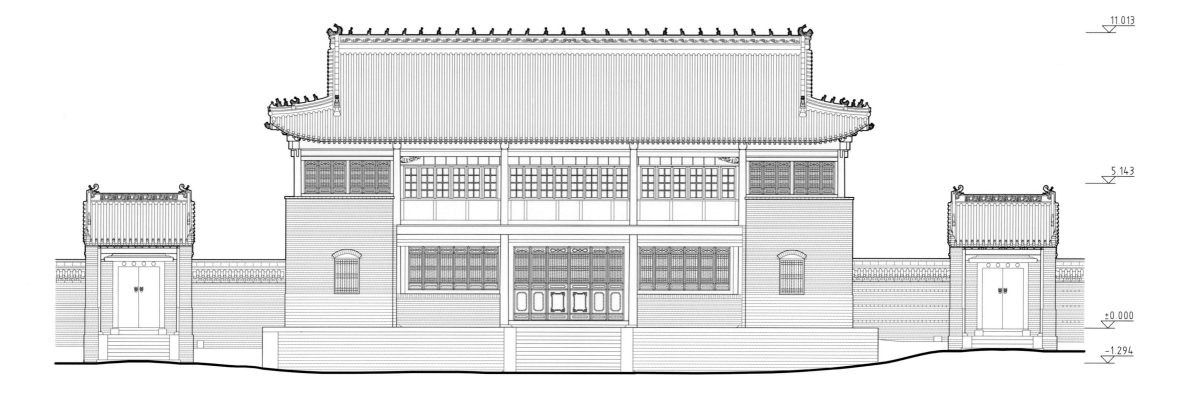

11.013

5.143

±0.000

-1.294

中岳庙崇圣门正立面图
Front elevation of Chongshengmen of Zhongyue Temple

0　　1　　2m

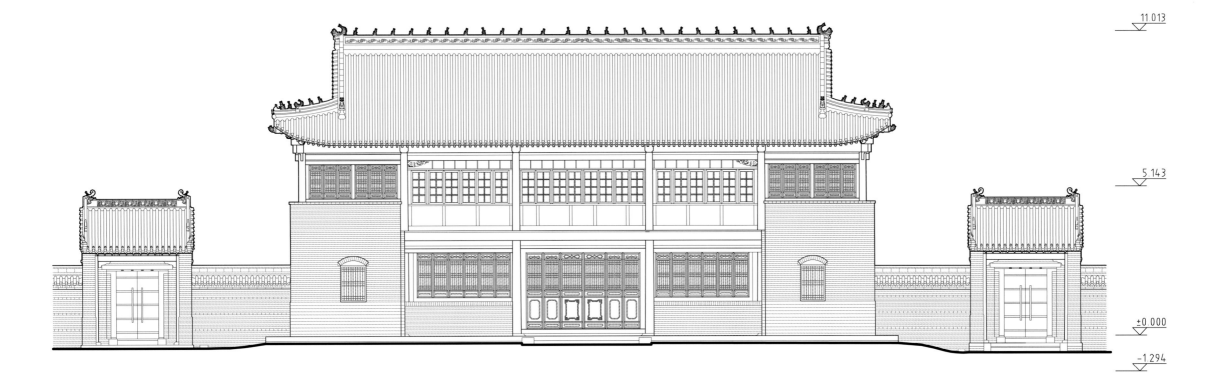

11 013

5 143

±0 000

-1 294

中岳庙崇圣门背立面图
Rear elevation of Chongshengmen of Zhongyue Temple

0 1 2m

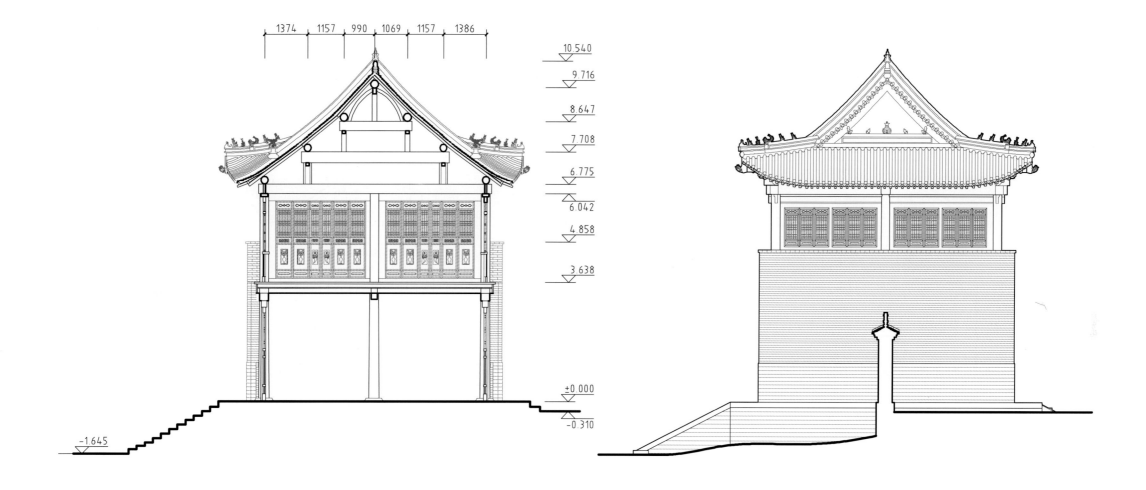

1374   1157   990   1069   1157   1386

10.540

9.716

8.647

7.708

6.775

6.042

4.858

3.638

±0.000

-0.310

-1.645

中岳庙崇圣门横剖面图
Cross-section of Chongshengmen of Zhongyue Temple

中岳庙崇圣门侧立面图
Side elevation of Chongshengmen of Zhongyue Temple

0   1   2m

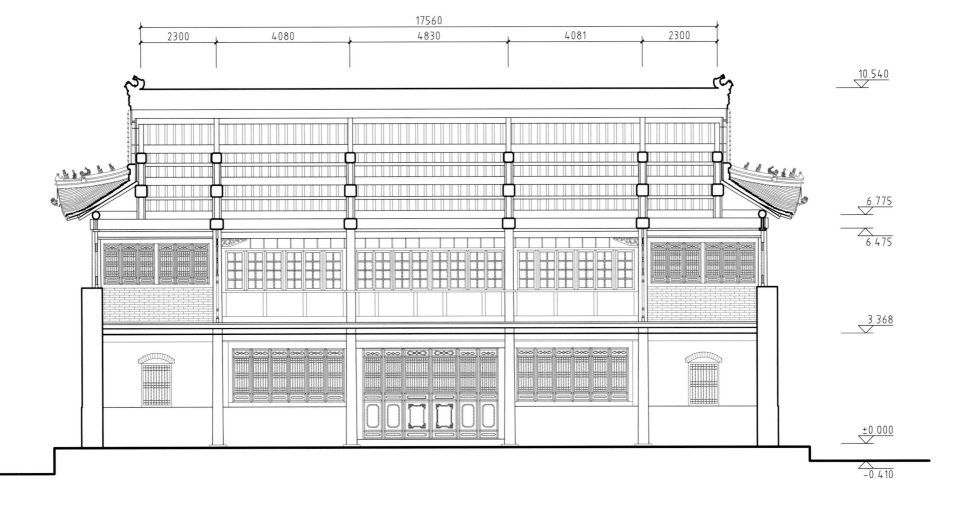

中岳庙崇圣门纵剖面图
Longitudinal section of Chongshengmen of Zhongyue Temple

0   1   2m

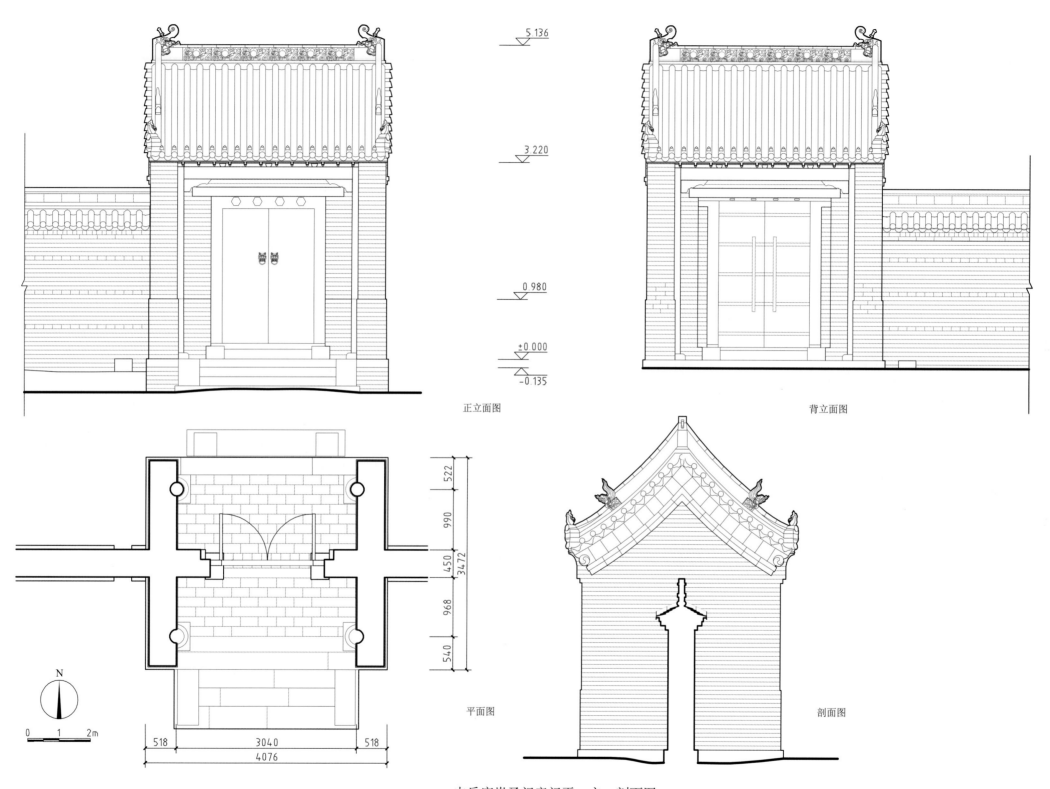

5.136

3.220

0.980

±0.000

−0.135

正立面图

背立面图

522

990

450

3472

968

540

N

0 1 2m

518 3040 518

4076

平面图

剖面图

中岳庙崇圣门旁门平、立、剖面图

Plan, elevation and section of *pangmen* of Chongshengmen, Zhongyue Temple

0 1 2m

二层隔扇门

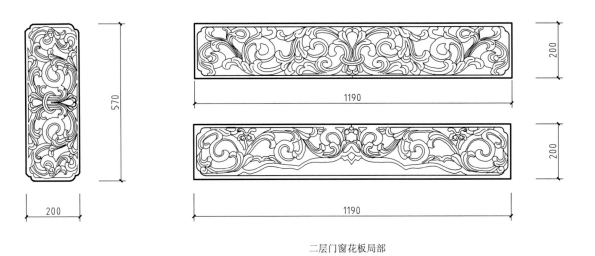

二层门窗花板局部

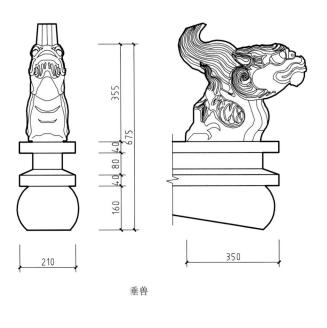

垂兽

中岳庙崇圣门构件大样图
Structural components of Chongshengmen of Zhongyue Temple

0 0.1 0.2m

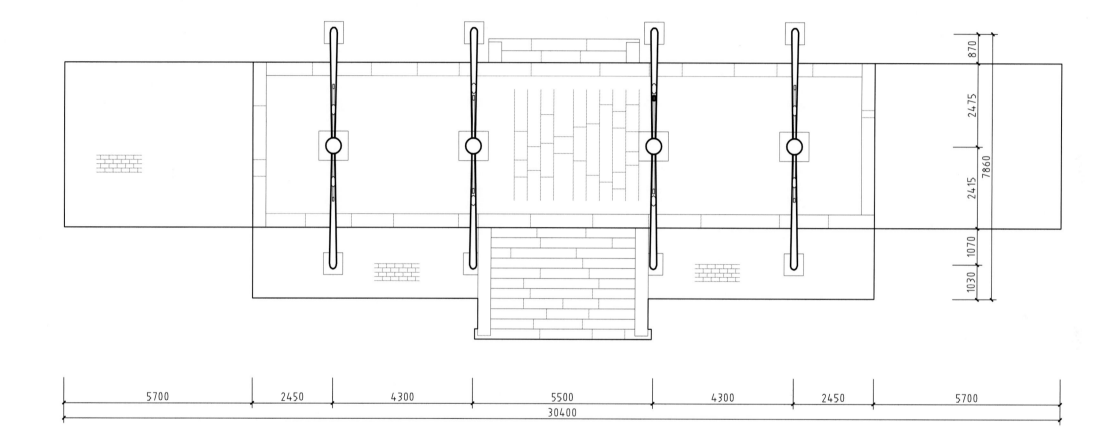

中岳庙配天作镇牌坊平面图
Plan of Peitian zuozhen *paifang* of Zhongyue Temple

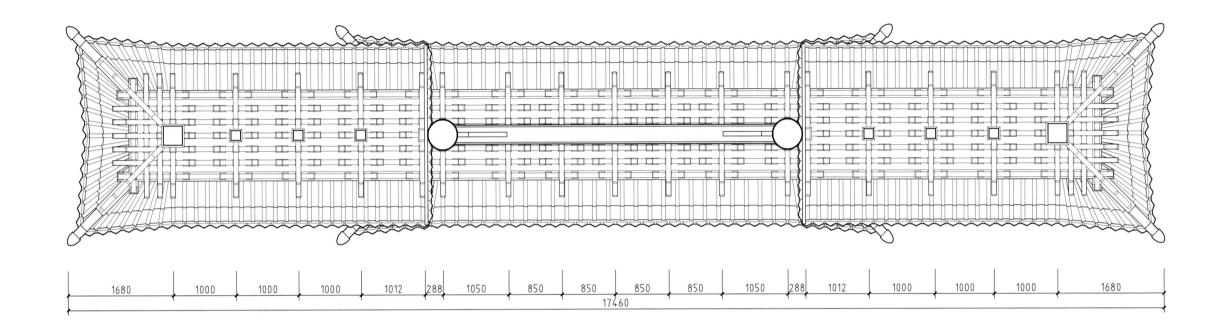

1680　1000　1000　1000　1012　288　1050　850　850　850　850　1050　288　1012　1000　1000　1000　1680

17460

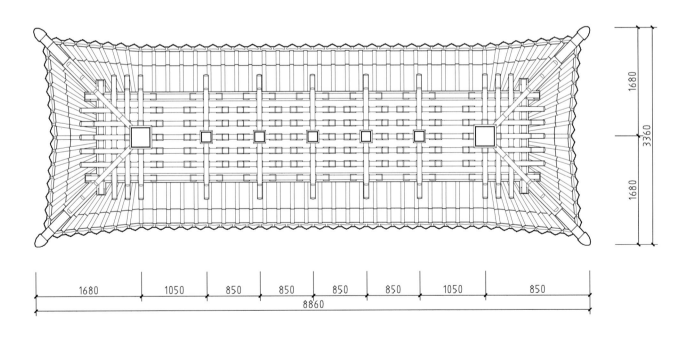

1680　1050　850　850　850　850　1050　850

8860

1680　3360　1680

中岳庙配天作镇牌坊梁架仰视平面图

Framework plan of Peitian zuozhen *paifang* of Zhongyue Temple as seen from below

0　1　2m

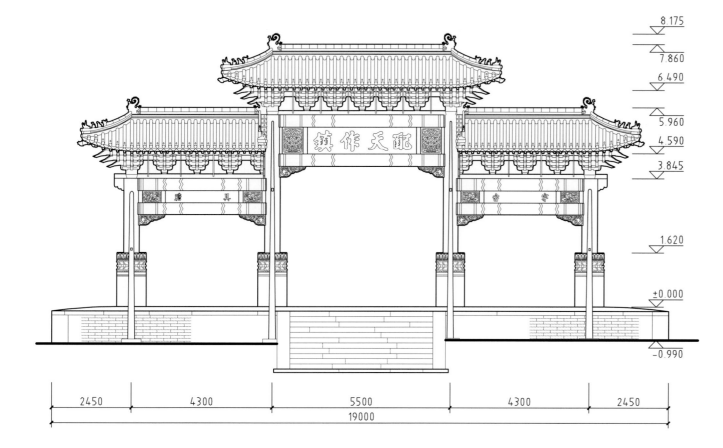

中岳庙配天作镇牌坊正立面图
Front elevation of Peitian zuozhen *paifang* of Zhongyue Temple

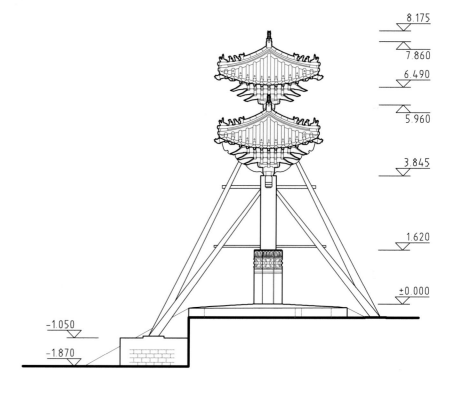

中岳庙配天作镇牌坊侧立面图
Side elevation of Peitian zuozhen *paifang* of Zhongyue Temple

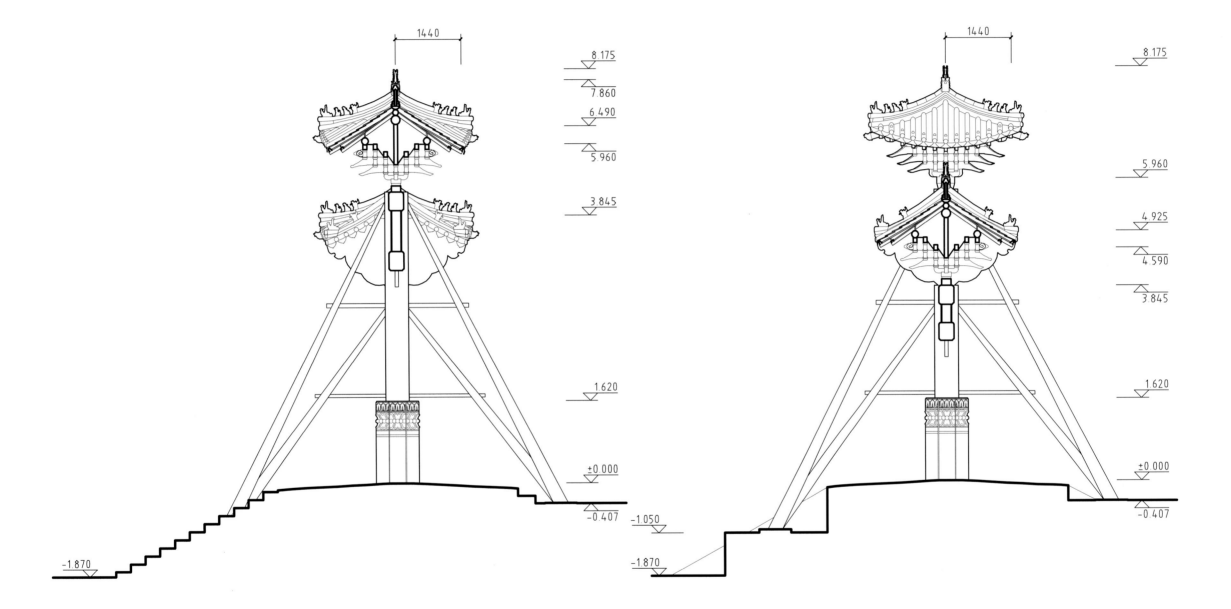

1440

8.175
7.860
6.490
5.960
3.845
1.620
±0.000
-0.407
-1.870

1440

8.175
5.960
4.925
4.590
3.845
1.620
±0.000
-0.407
-1.050
-1.870

0   1   2m

中岳庙配天作镇牌坊明间横剖面图
Cross-section of central-bay of Peitian zuozhen *paifang*, Zhongyue Temple

中岳庙配天作镇牌坊次间横剖面图
Cross-section of side-bay of Peitian zuozhen *paifang*, Zhongyue Temple

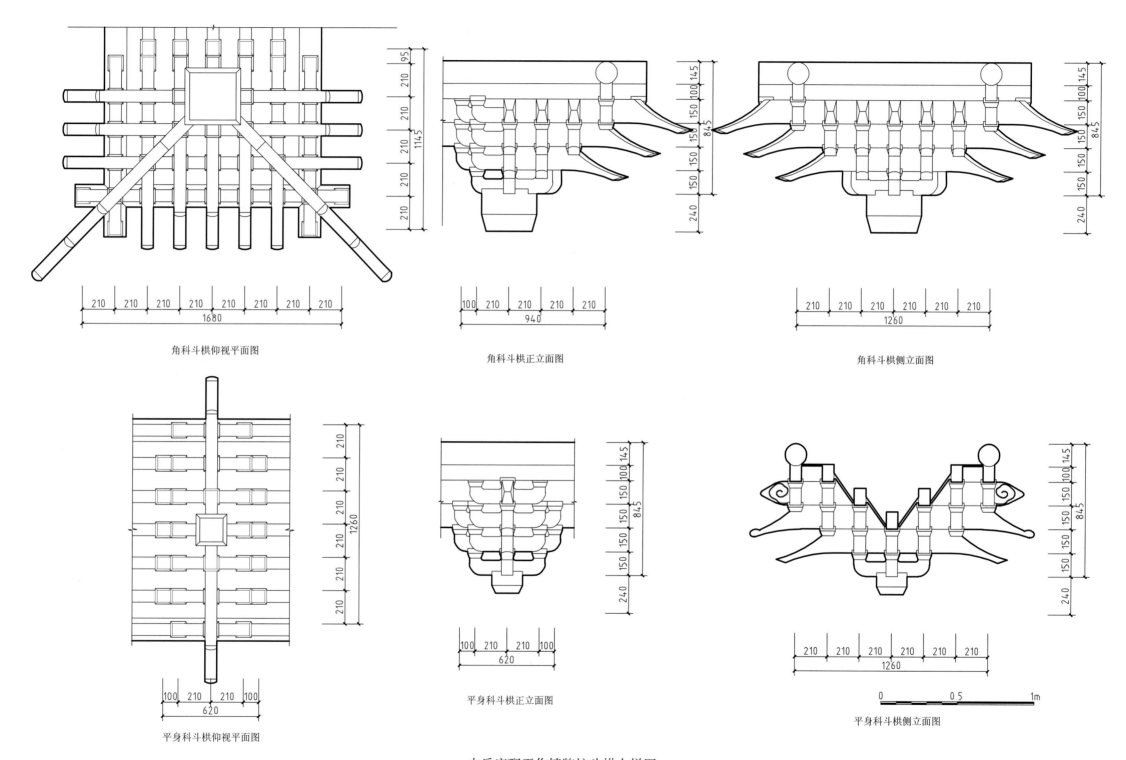

角科斗栱仰视平面图

角科斗栱正立面图

角科斗栱侧立面图

平身科斗栱仰视平面图

平身科斗栱正立面图

平身科斗栱侧立面图

中岳庙配天作镇牌坊斗栱大样图
Bracket set of Peitian zuozhen *paifang* of Zhongyue Temple

夹杆石石作

明间花板

次间花板

雀替木作

瓦当

滴水

角梁套兽瓦作

戗脊走兽

戗兽

正吻

中岳庙配天作镇牌坊构件大样图
Structural components of Peitian zuozhen *paifang* of Zhongyue Temple

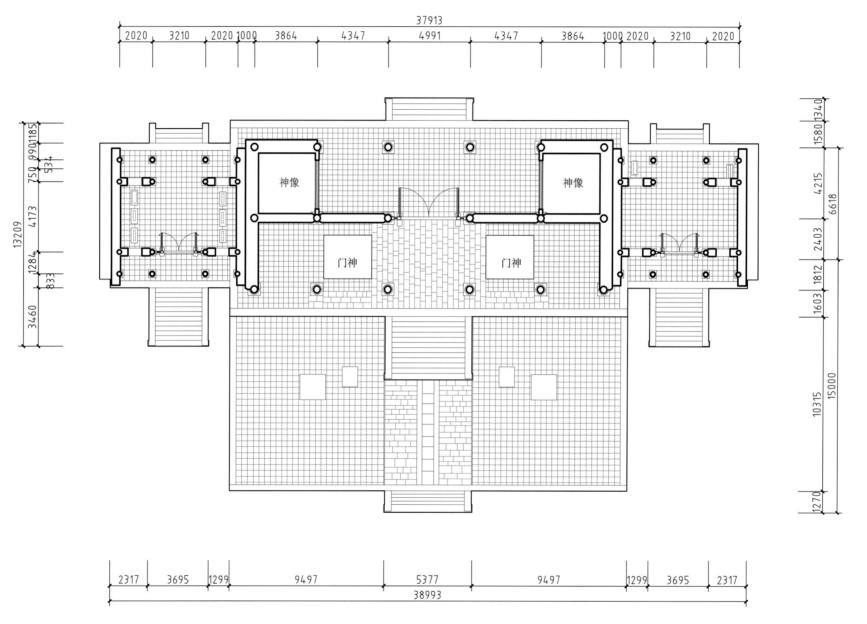

中岳庙峻极门总平面图
Site plan of Junjimen of Zhongyue Temple

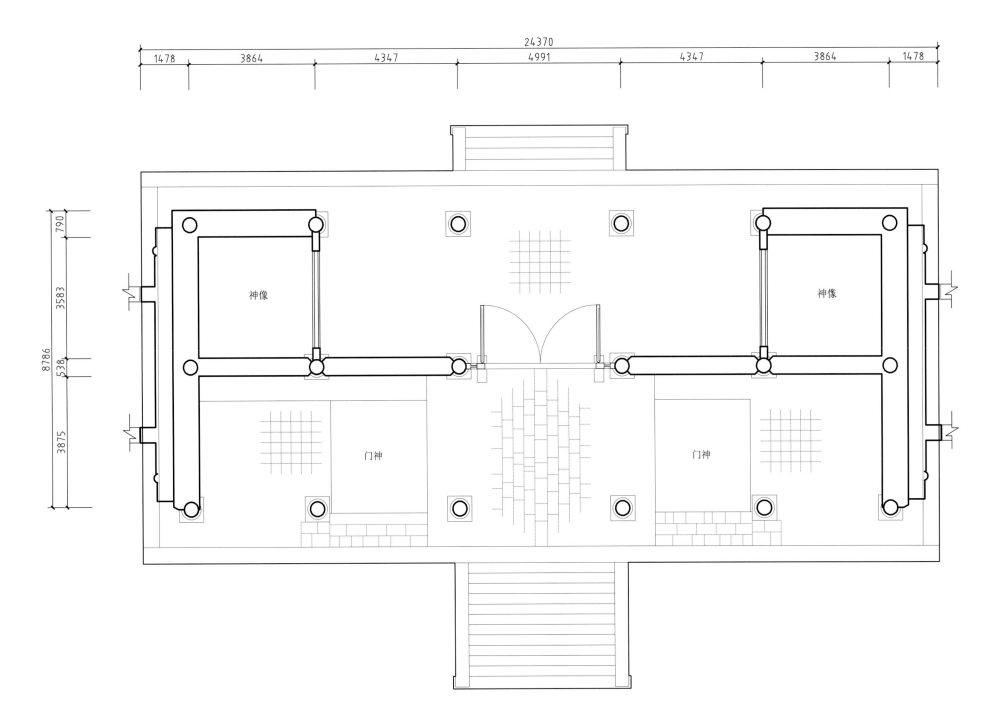

中岳庙峻极门首层平面图
Plan of first floor of Junjimen, Zhongyue Temple

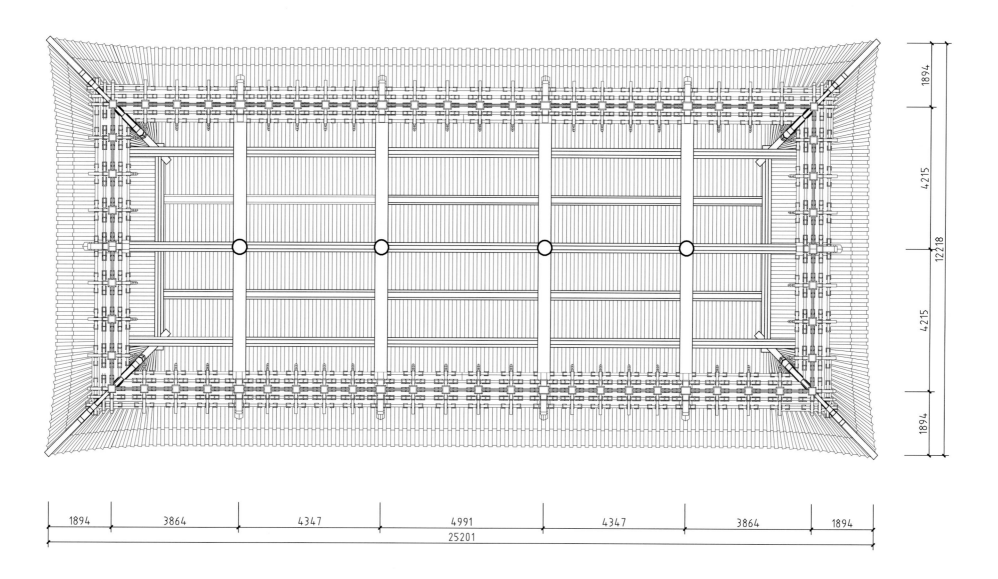

中岳庙峻极门梁架仰视平面图
Framework plan of Junjimen of Zhongyue Temple as seen from below

0 1 2m

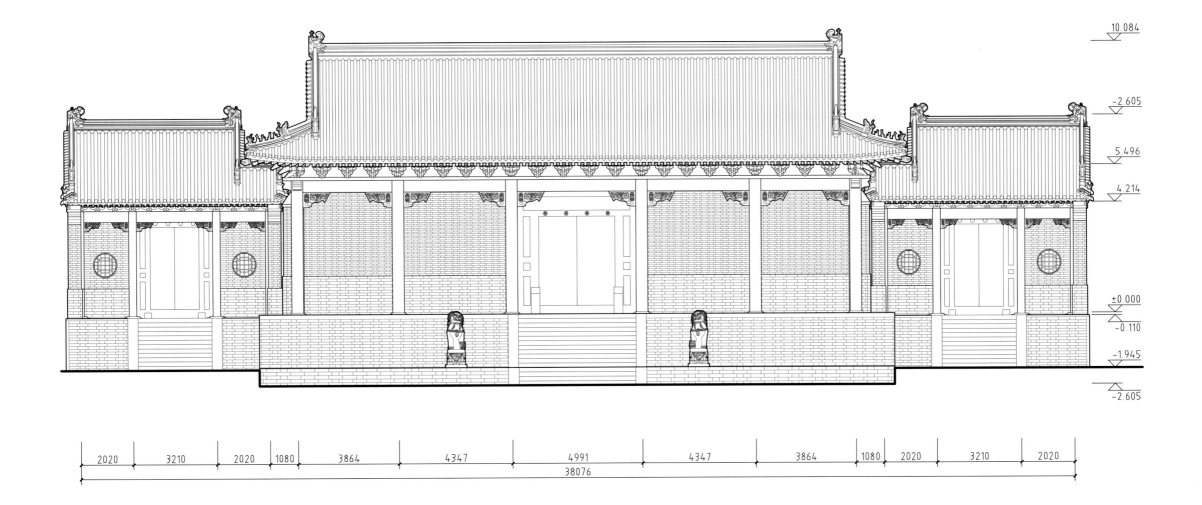

中岳庙峻极门总正立面图
Site elevation of front of Junjimen, Zhongyue Temple

10.084

-2.605

5.496

4.214

±0.000
-0.110

-1.945

-2.605

2020　3210　2020　1080　3864　4347　4991　4347　3864　1080　2020　3210　2020
38076

0　1　2m

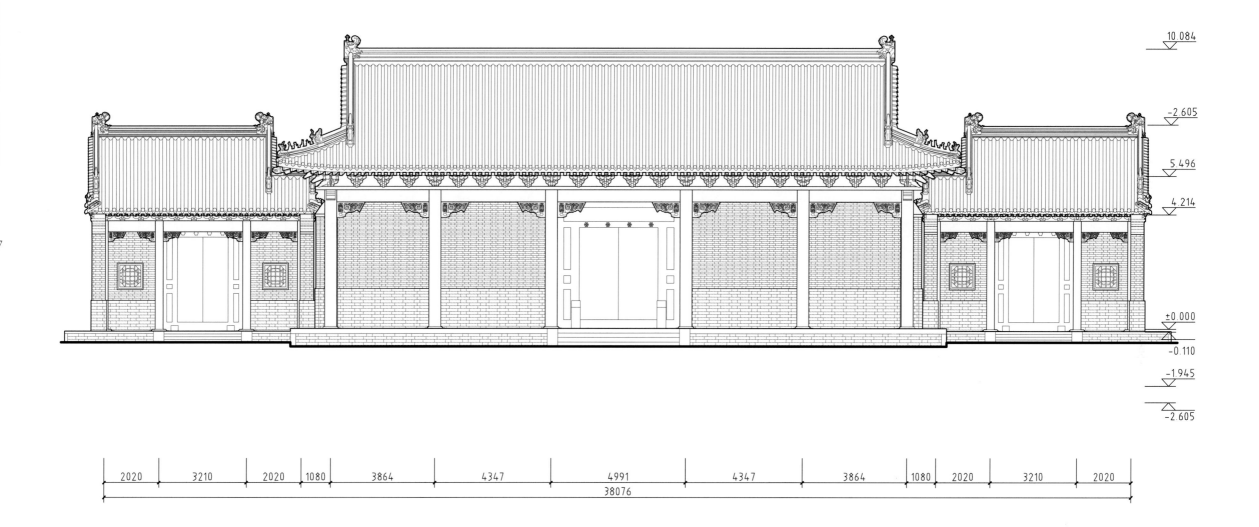

10.084

-2.605

5.496

4.214

±0.000

-0.110

-1.945

-2.605

| 2020 | 3210 | 2020 | 1080 | 3864 | 4347 | 4991 | 4347 | 3864 | 1080 | 2020 | 3210 | 2020 |

38076

中岳庙峻极门总背立面图
Site elevation of rear of Junjimen, Zhongyue Temple

0   1   2m

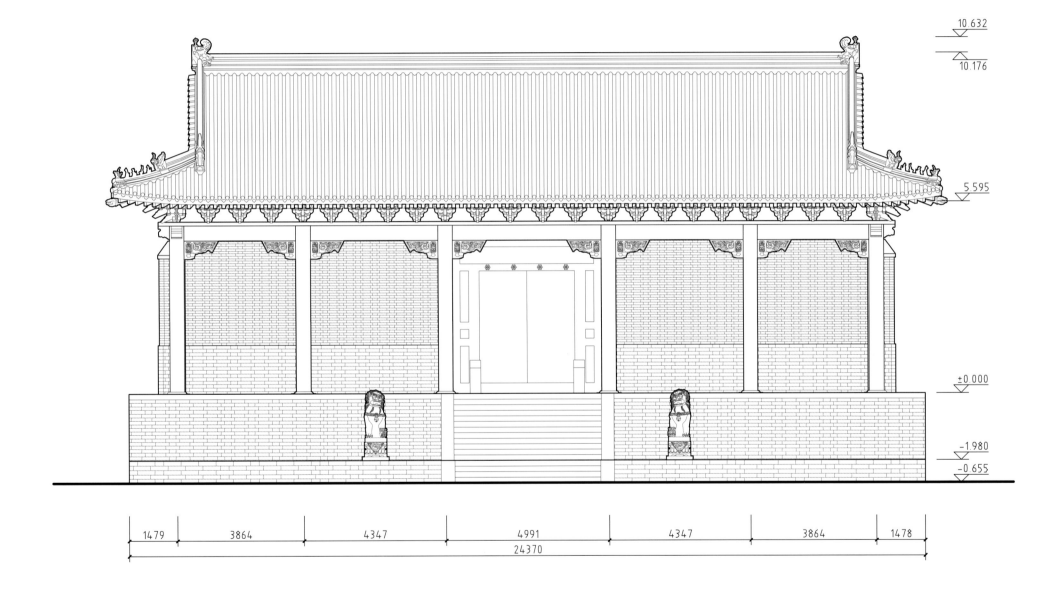

中岳庙峻极门正立面图
Front elevation of Junjimen, Zhongyue Temple

0　1　2m

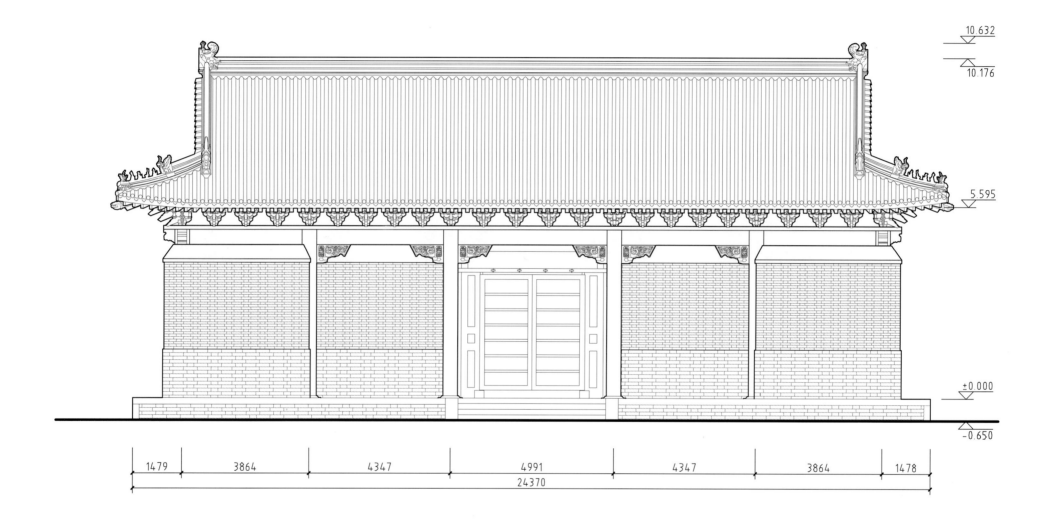

10.632

10.176

5.595

±0.000

-0.650

| 1479 | 3864 | 4347 | 4991 | 4347 | 3864 | 1478 |

24370

中岳庙峻极门背立面图
Rear elevation of Junjimen, Zhongyue Temple

0   1   2m

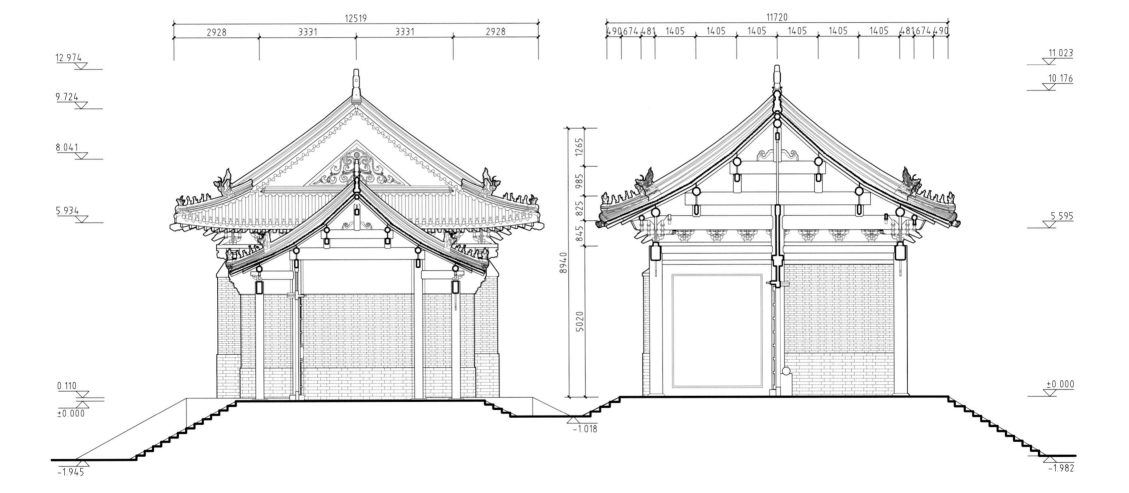

中岳庙峻极门侧立面图
Side elevation of Junjimen of Zhongyue Temple

中岳庙峻极门横剖面图
Cross-section of Junjimen of Zhongyue Temple

0 1 2m

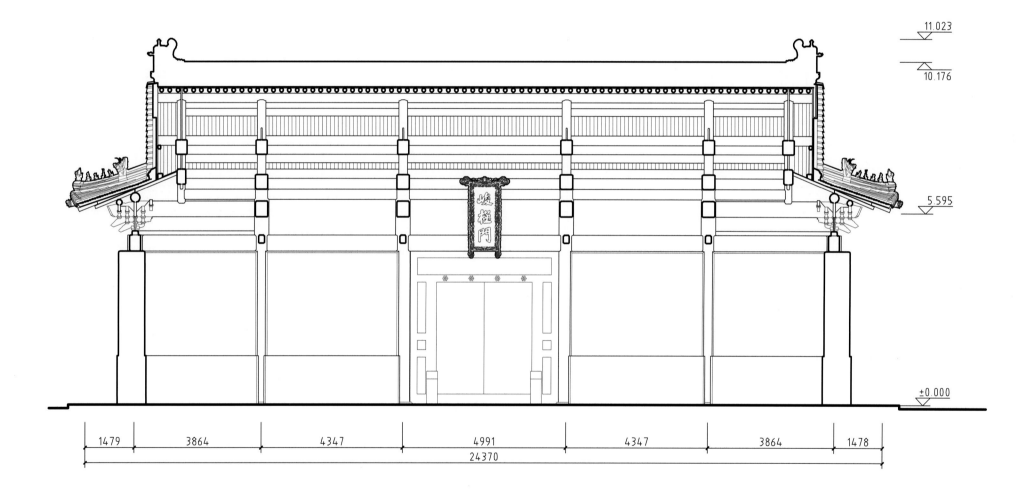

11.023

10.176

5.595

±0.000

| 1479 | 3864 | 4347 | 4991 | 4347 | 3864 | 1478 |

24370

中岳庙峻极门纵剖面图

Longitudinal section of Junjimen of Zhongyue Temple

0　1　2m

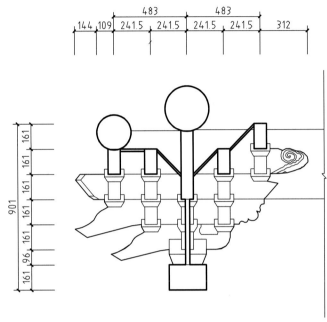

平身科斗栱剖面图

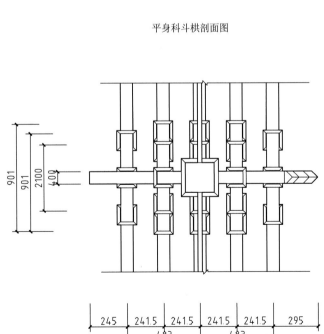

平身科斗栱平面图

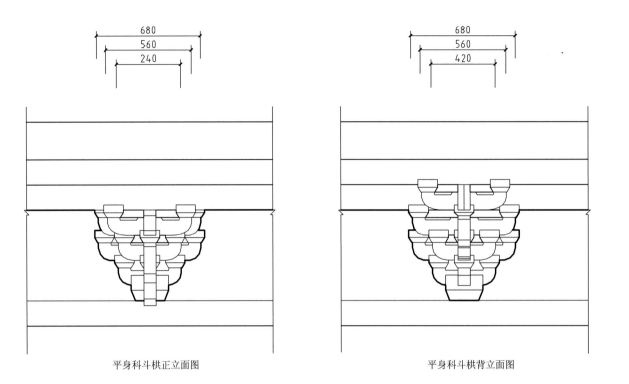

平身科斗栱正立面图

平身科斗栱背立面图

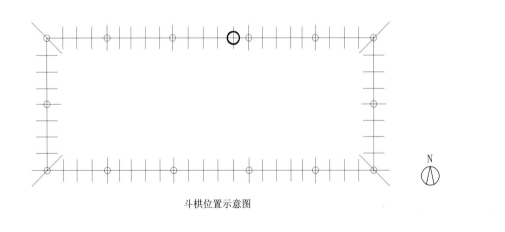

斗栱位置示意图

中岳庙峻极门斗栱大样图（一）
Bracket set of Junjimen of Zhongyue Temple Ⅰ

0   0.5   1m

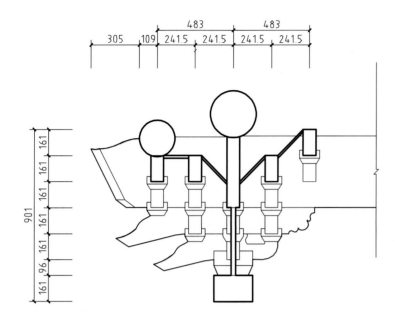

柱头科斗栱剖面图

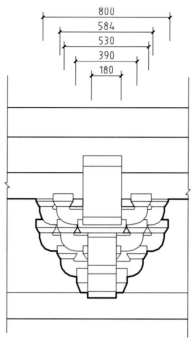

柱头科斗栱正立面图

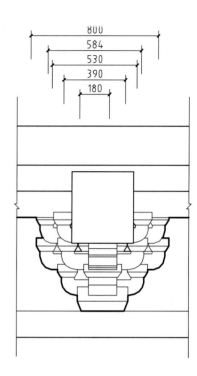

柱头科斗栱背立面图

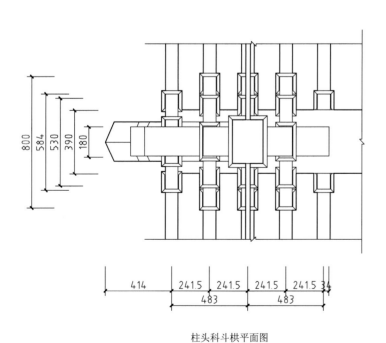

柱头科斗栱平面图

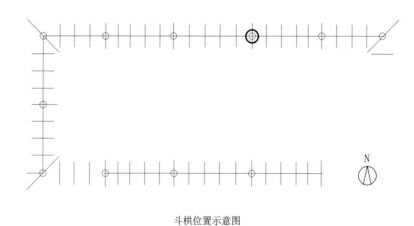

斗栱位置示意图

中岳庙峻极门斗栱大样图（二）
Bracket set of Junjimen of Zhongyue Temple Ⅱ

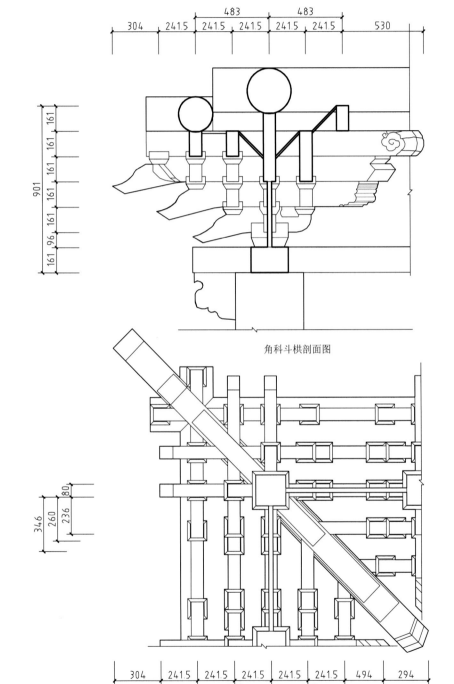

角科斗栱剖面图

角科斗栱平面图

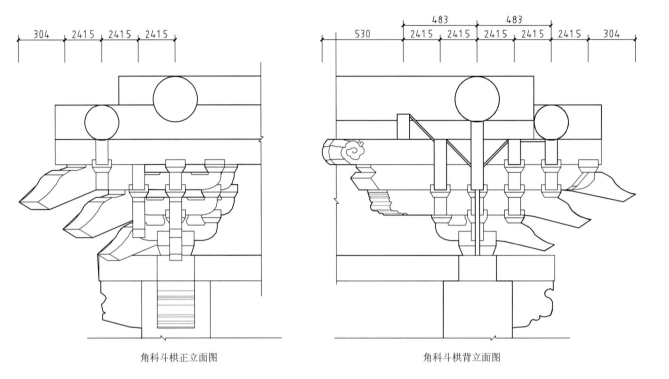

角科斗栱正立面图

角科斗栱背立面图

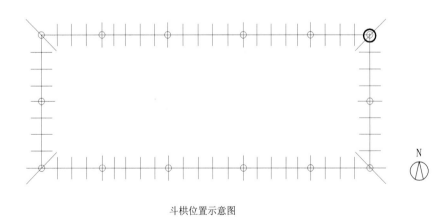

斗栱位置示意图

N

中岳庙峻极门斗栱大样图（三）

Bracket set of Junjimen of Zhongyue Temple III

0   0.5   1m

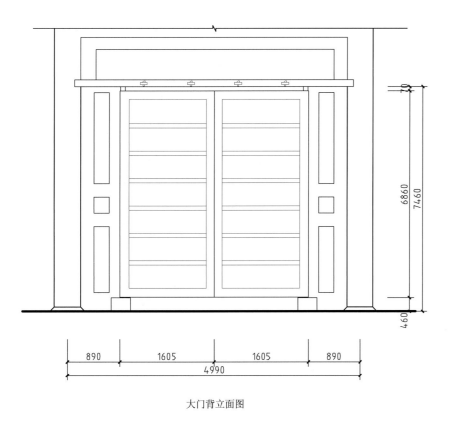

大门背立面图

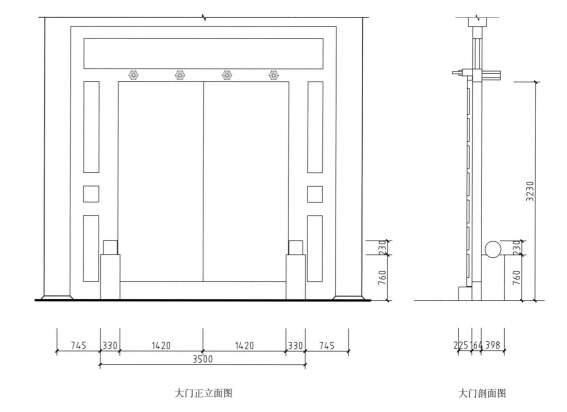

大门正立面图

大门剖面图

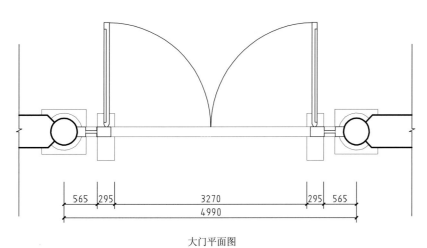

大门平面图

中岳庙峻极门构件大样图（一）
Structural components of Junjimen of Zhongyue Temple Ⅰ

0 0.5 1m

100

550

550

1050

325　520　205

1050

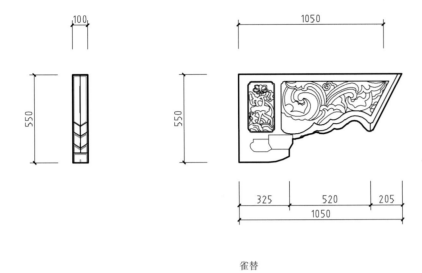

雀替

1520

2420

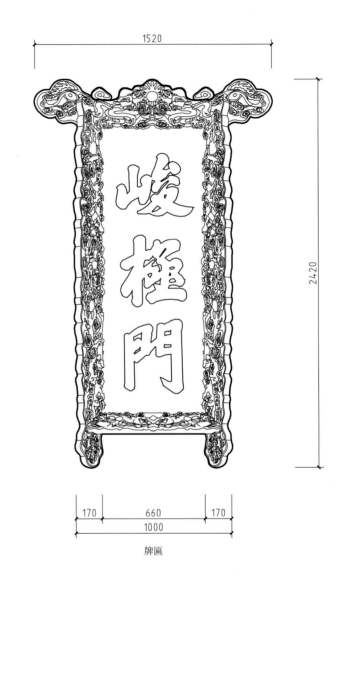

170　660　170

1000

牌匾

800

805

137

220

343

1120

480

80

90

1120

275

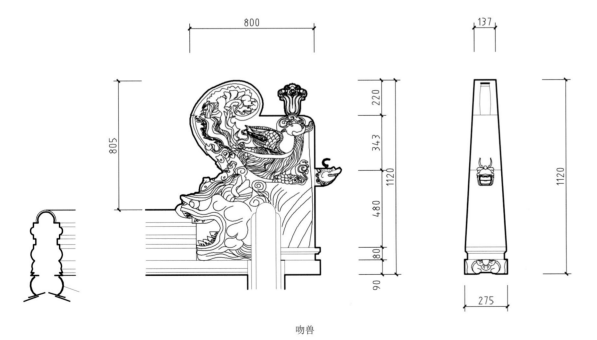

吻兽

中岳庙峻极门构件大样图（二）
Structural components of Junjimen of Zhongyue Temple Ⅱ

0　　0.5　　1m

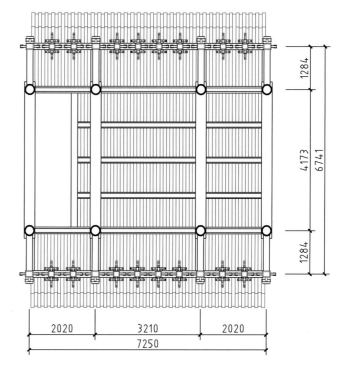

中岳庙峻极门东角门平面图
Plan of Dongjiaomen of Junjimen, Zhongyue Temple

中岳庙峻极门东角门梁架仰视平面图
Framework plan of Dongjiaomen of Zhongyue Temple Junjimen as seen from below

N

0 1 2m

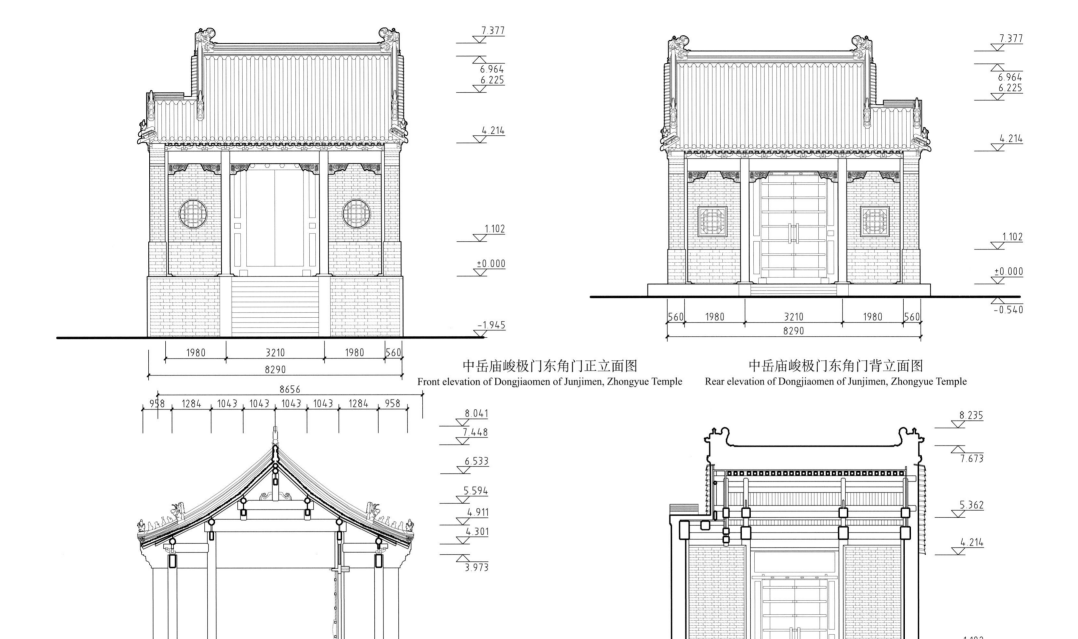

中岳庙峻极门东角门正立面图
Front elevation of Dongjiaomen of Junjimen, Zhongyue Temple

中岳庙峻极门东角门背立面图
Rear elevation of Dongjiaomen of Junjimen, Zhongyue Temple

中岳庙峻极门东角门横剖面图
Cross-section of Dongjiaomen of Junjimen, Zhongyue Temple

中岳庙峻极门东角门纵剖面图
Longitudinal section of Dongjiaomen of Junjimen, Zhongyue Temple

0   1   2m

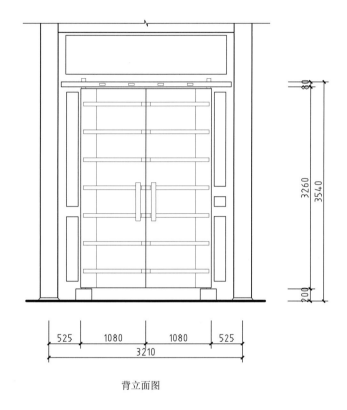

背立面图

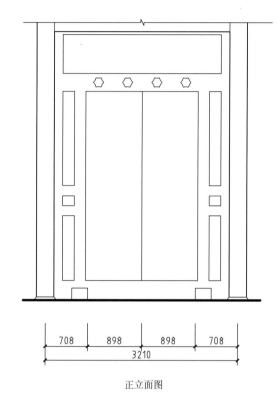

正立面图

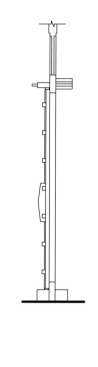

剖面图

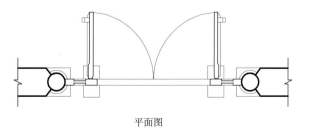

平面图

中岳庙峻极门东角门大门大样图
Main gate of Dongjiaomen of Junjimen, Zhongyue Temple

0   0.5   1m

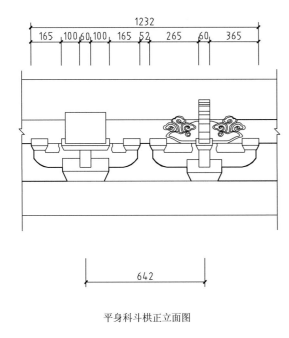

平身科斗栱正立面图

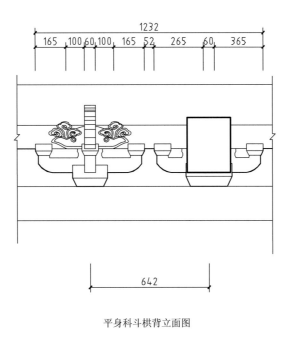

平身科斗栱背立面图

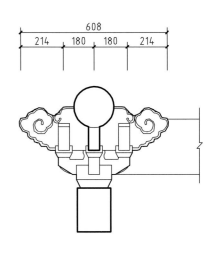

平身科斗栱剖面图

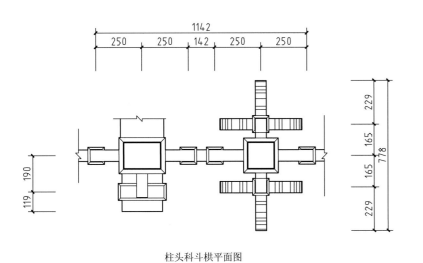

柱头科斗栱平面图

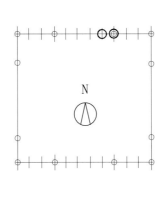

斗栱位置示意图

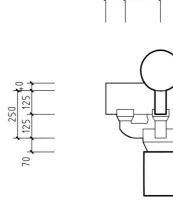

柱头科斗栱剖面图

中岳庙峻极门东角门斗栱大样图
Bracket set of Dongjiaomen of Junjimen, Zhongyue Temple

0   0.2   0.5m

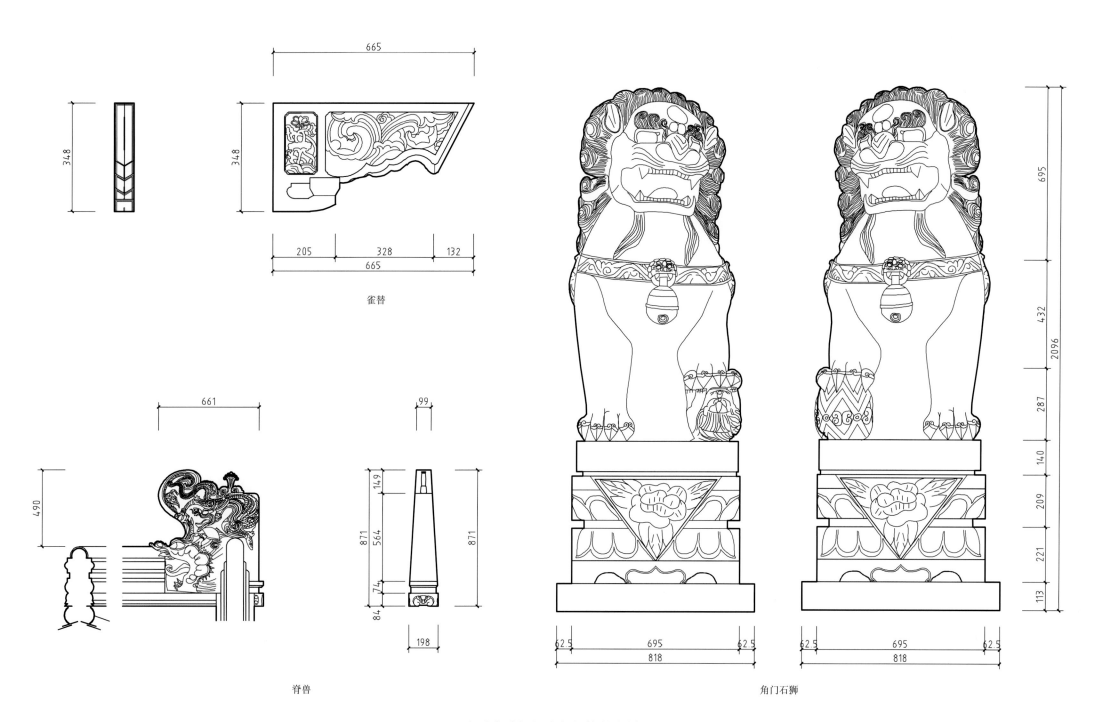

665

348

348

205　328　132
665

雀替

661

490

99

149
871
564
871
74
84
198

脊兽

695

4.32

2096

287

140

209

221

113

62.5　695　62.5
818

62.5　695　62.5
818

角门石狮

中岳庙峻极门东角门构件大样图
Structural components of Dongjiaomen of Junjimen, Zhongyue Temple

0　0.2　0.5m

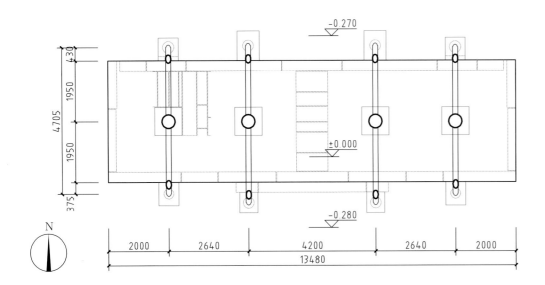

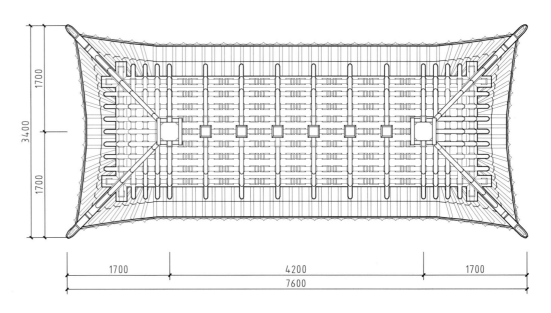

中岳庙嵩高峻极牌坊平面图
Plan of Songgao junji *paifang* of Zhongyue Temple

中岳庙嵩高峻极牌坊梁架仰视平面图（一）
Framework plan of Songgao junji *paifang* of Zhongyue Temple as seen from below Ⅰ

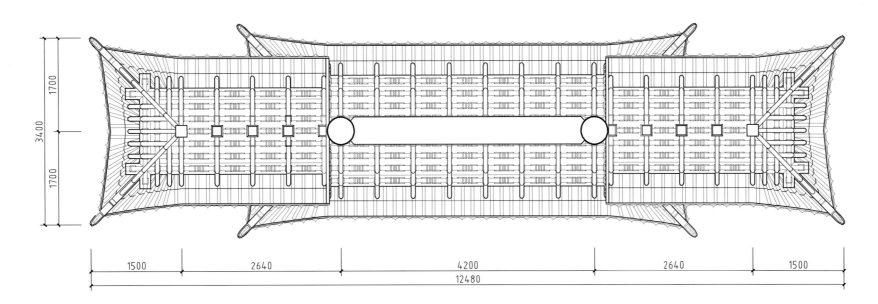

中岳庙嵩高峻极牌坊梁架仰视平面图（二）
Framework plan of Songgao junji *paifang* of Zhongyue Temple as seen from below Ⅱ

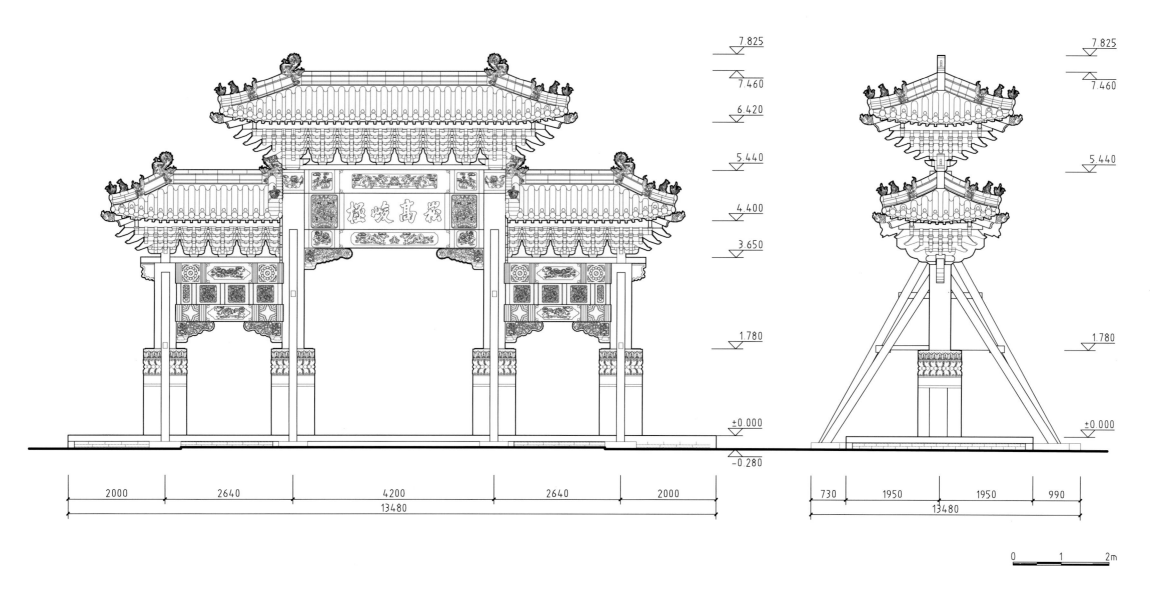

中岳庙嵩高峻极牌坊正立面图
Front elevation of Songgao junji *paifang* of Zhongyue Temple

中岳庙嵩高峻极牌坊侧立面图
Side elevation of Songgao junji *paifang* of Zhongyue Temple

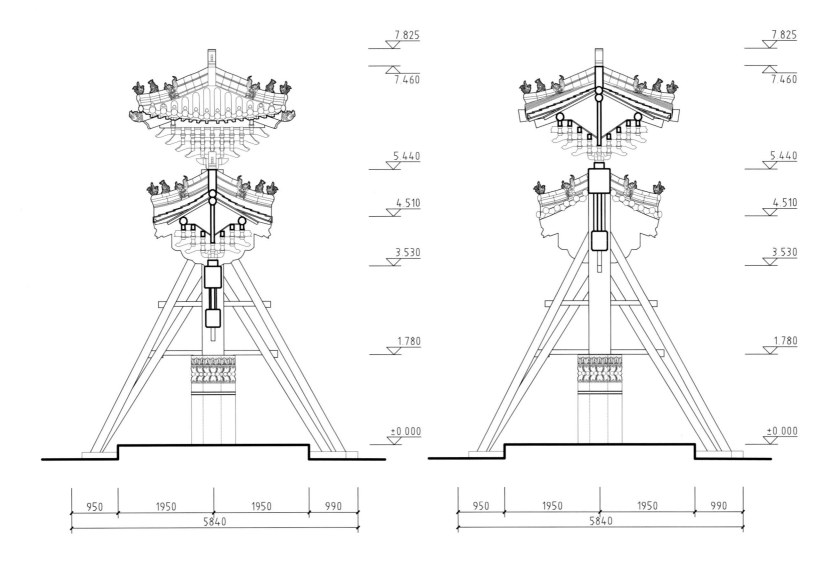

中岳庙嵩高峻极牌坊次间剖面图
Section of side-bay of Songgao junji *paifang*, Zhongyue Temple

中岳庙嵩高峻极牌坊明间剖面图
Section of central-bay of Songgao junji *paifang*, Zhongyue Temple

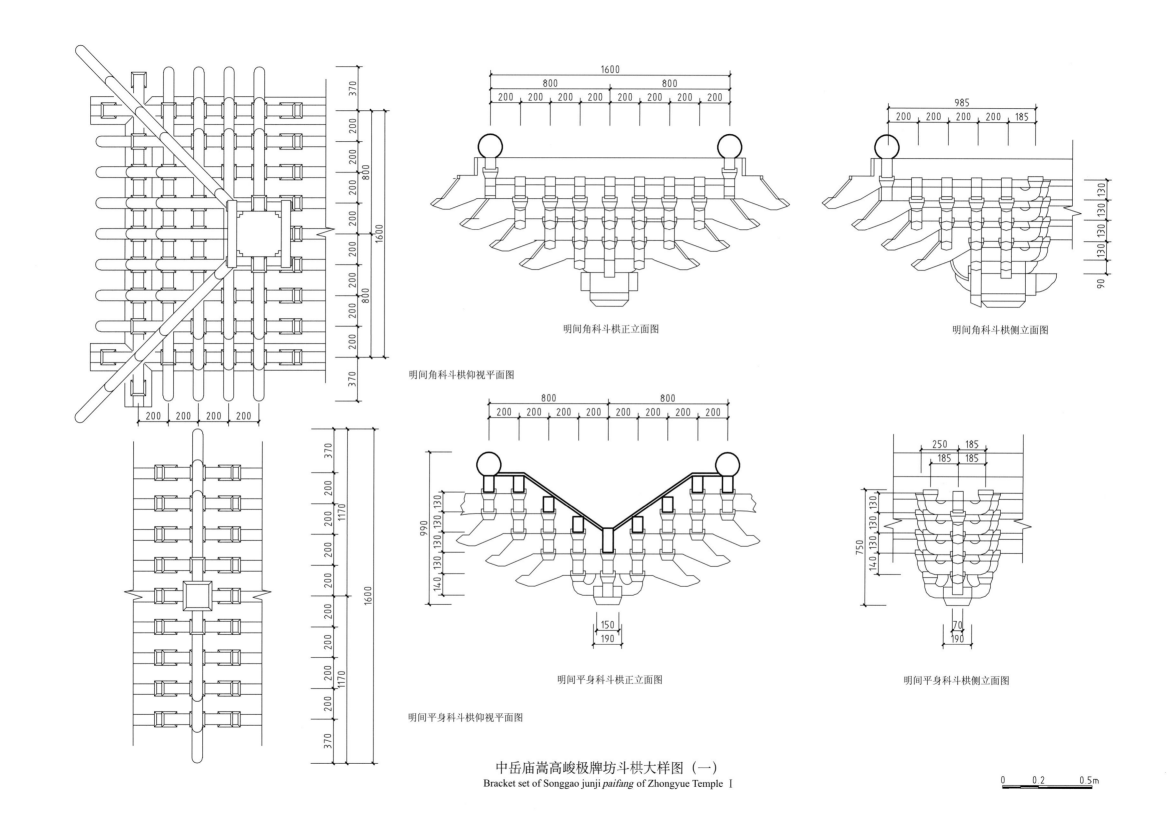

明间角科斗栱正立面图

明间角科斗栱侧立面图

明间角科斗栱仰视平面图

明间平身科斗栱正立面图

明间平身科斗栱侧立面图

明间平身科斗栱仰视平面图

中岳庙嵩高峻极牌坊斗栱大样图 (一)
Bracket set of Songgao junji *paifang* of Zhongyue Temple I

0   0.2   0.5m

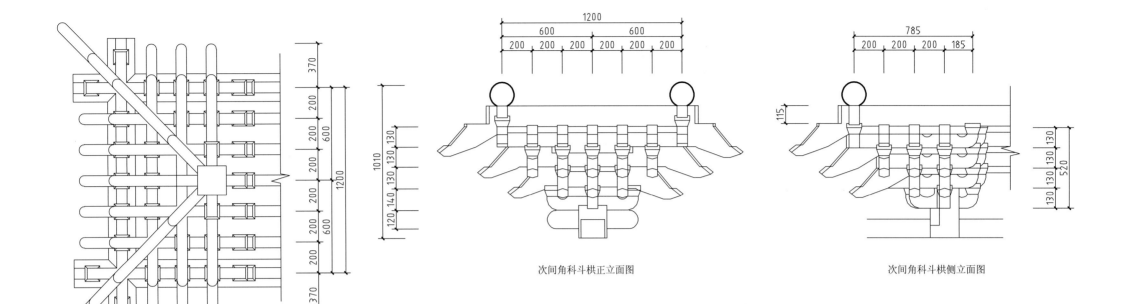

次间角科斗栱仰视平面图

次间角科斗栱正立面图

次间角科斗栱侧立面图

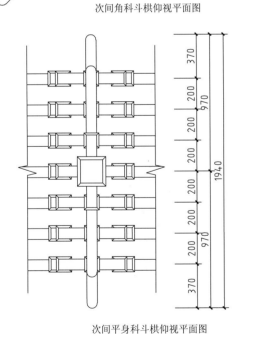

次间平身科斗栱仰视平面图

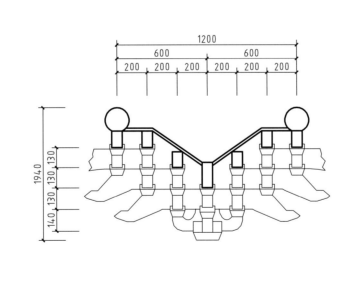

次间平身科斗栱正立面图

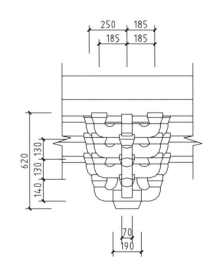

次间平身科斗栱侧立面图

中岳庙嵩高峻极牌坊斗栱大样图（二）

Bracket set of Songgao junji *paifang* of Zhongyue Temple Ⅱ

0 0.2 0.5m

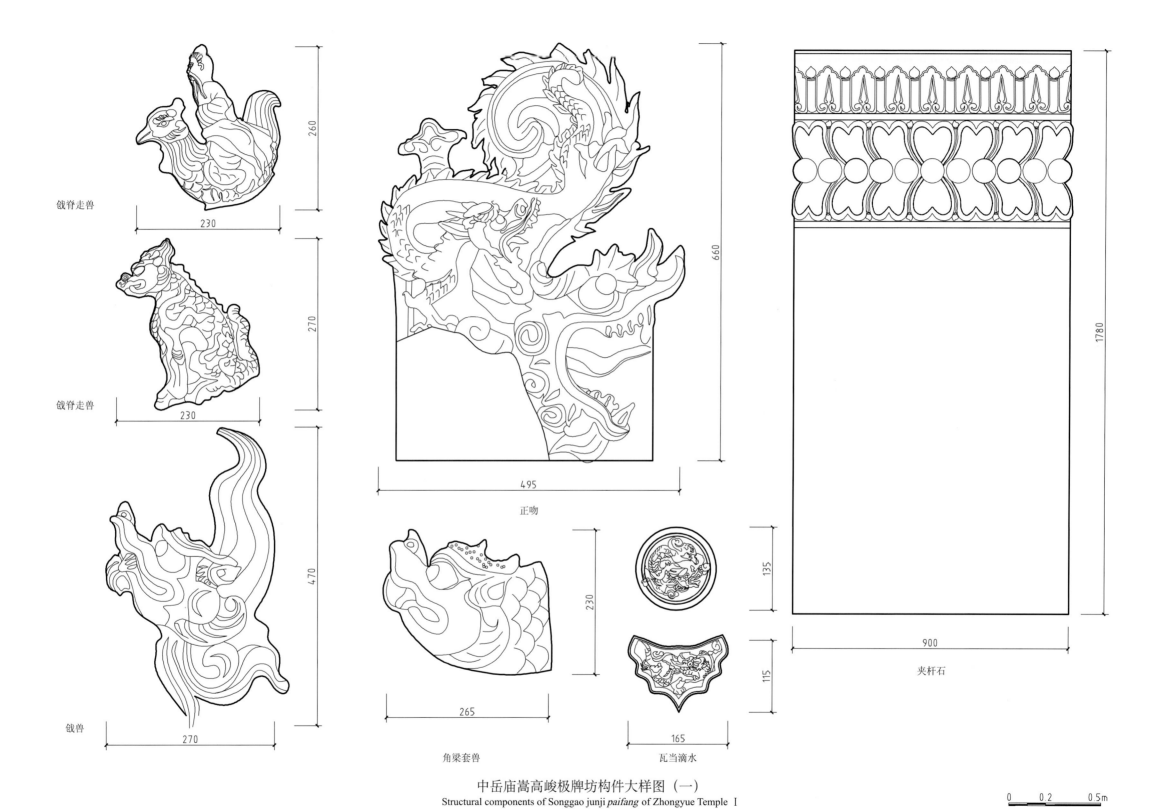

戗脊走兽

戗脊走兽

戗兽

正吻

角梁套兽

瓦当滴水

夹杆石

260

270

470

230

230

270

660

495

230

265

135

115

165

1780

900

中岳庙嵩高峻极牌坊构件大样图（一）
Structural components of Songgao junji *paifang* of Zhongyue Temple I

0   0.2   0.5m

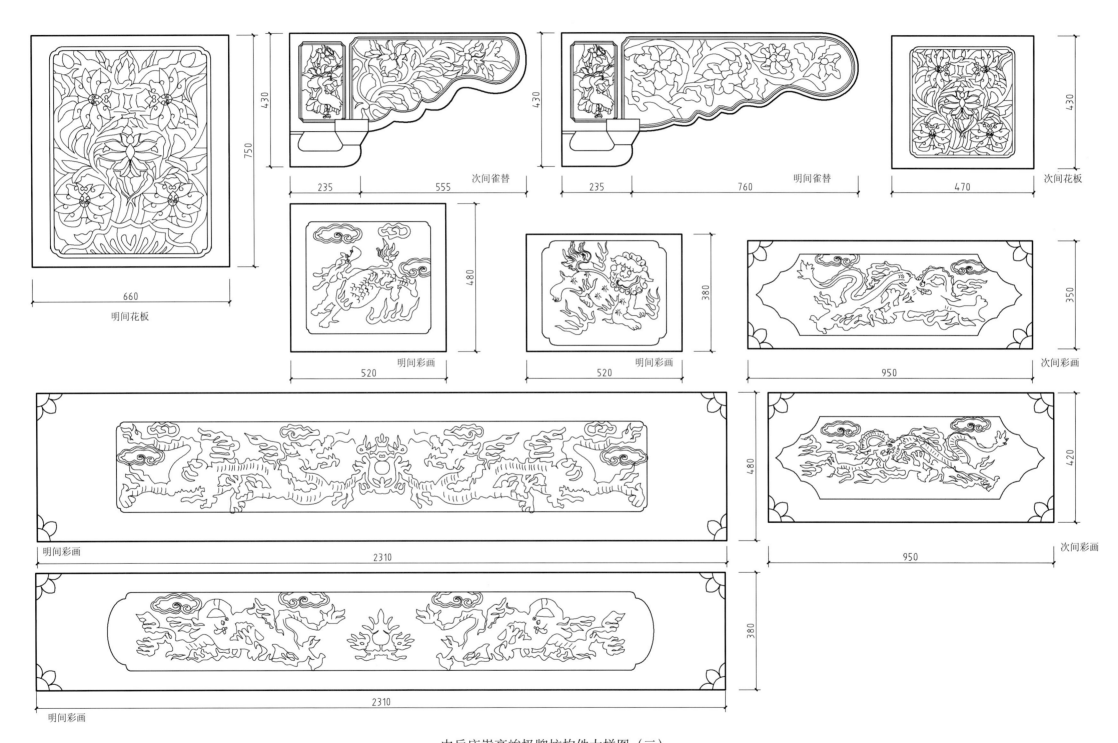

430

750

次间雀替

235   555

430

明间雀替

235   760

430

次间花板

470

660

明间花板

480

明间彩画

520

380

明间彩画

520

350

次间彩画

950

480

明间彩画

2310

420

次间彩画

950

380

明间彩画

2310

中岳庙嵩高峻极牌坊构件大样图（二）
Structural components of Songgao junji *paifang* of Zhongyue Temple Ⅱ

0   0.2   0.5m

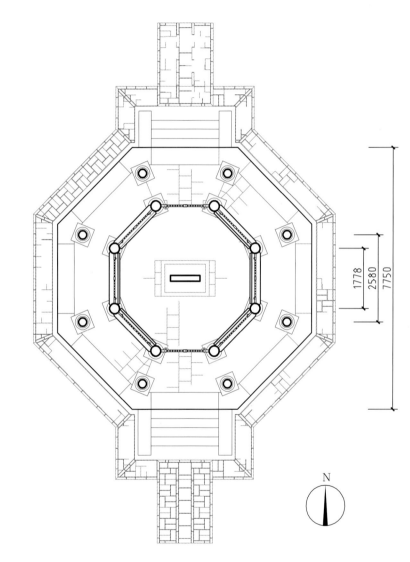

中岳庙御碑亭平面图
Plan of Yubeiting of Zhongyue Temple

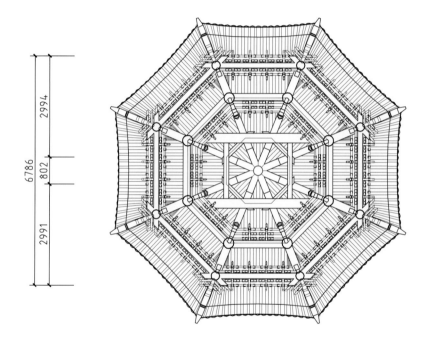

中岳庙御碑亭梁架仰视平面图
Framework plan of Yubeiting of Zhongyue Temple as seen from below

0 1 2m

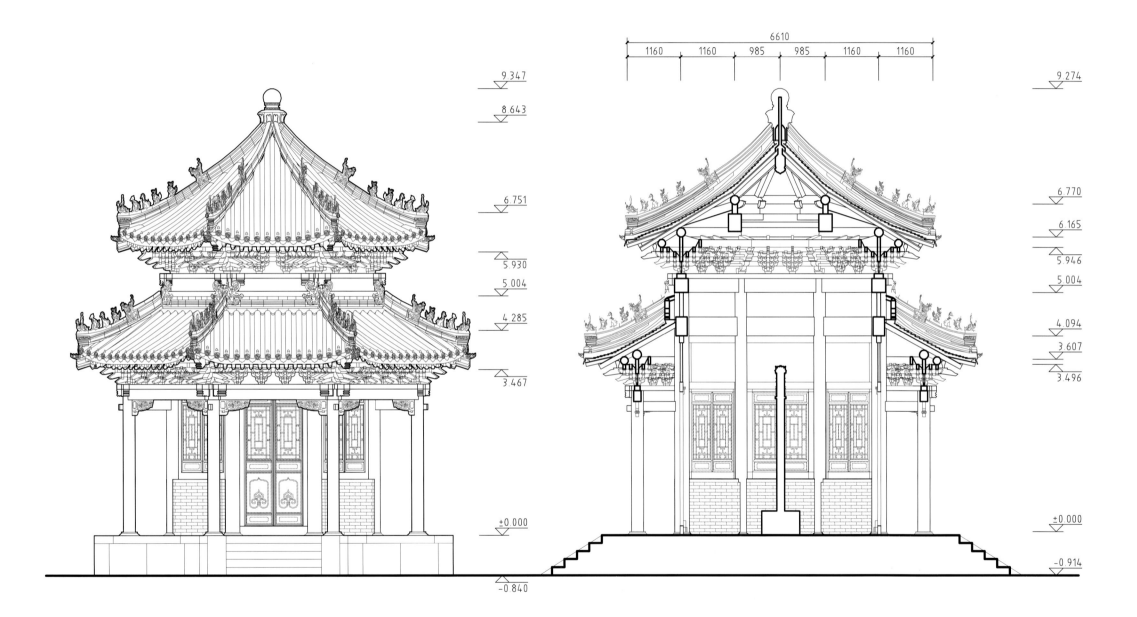

中岳庙御碑亭立面图
Elevation of Yubeiting of Zhongyue Temple

中岳庙御碑亭剖面图
Section of Yubeiting of Zhongyue Temple

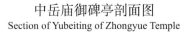

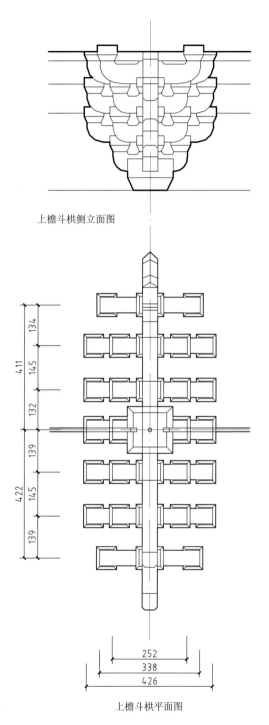

上檐斗栱侧立面图

上檐斗栱平面图

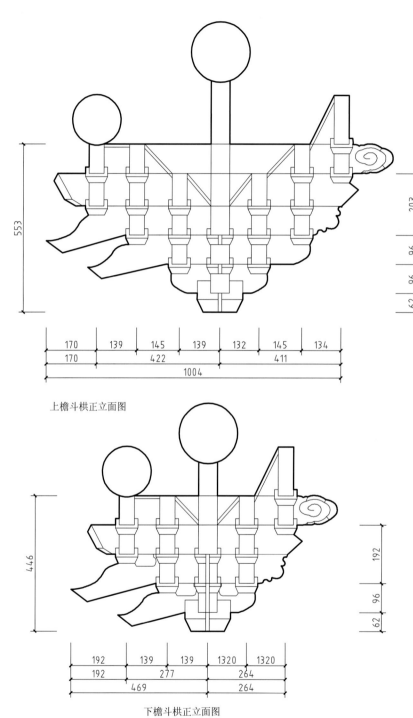

上檐斗栱正立面图

下檐斗栱正立面图

下檐斗栱平面图

下檐斗栱侧立面图

中岳庙御碑亭斗栱大样图

Bracket set of Yubeiting of Zhongyue Temple

0   01 0.2m

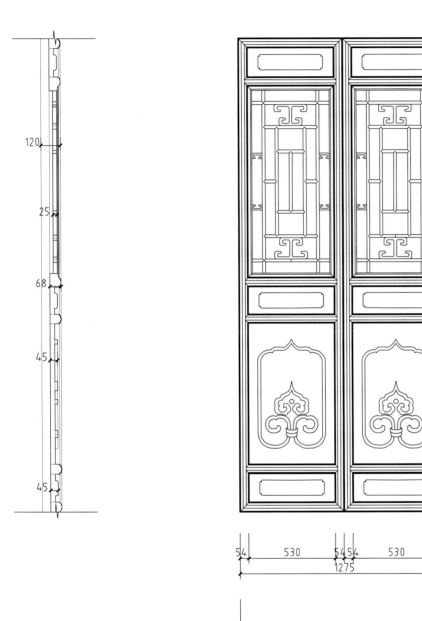

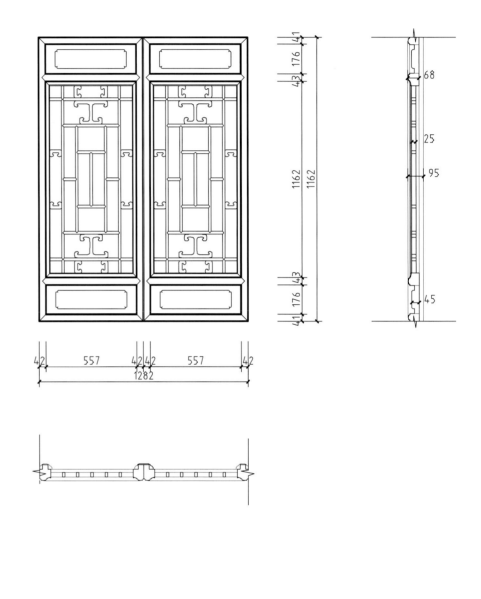

中岳庙御碑亭门窗大样图
Doors and windows of Yubeiting of Zhongyue Temple

0   0.5   1m

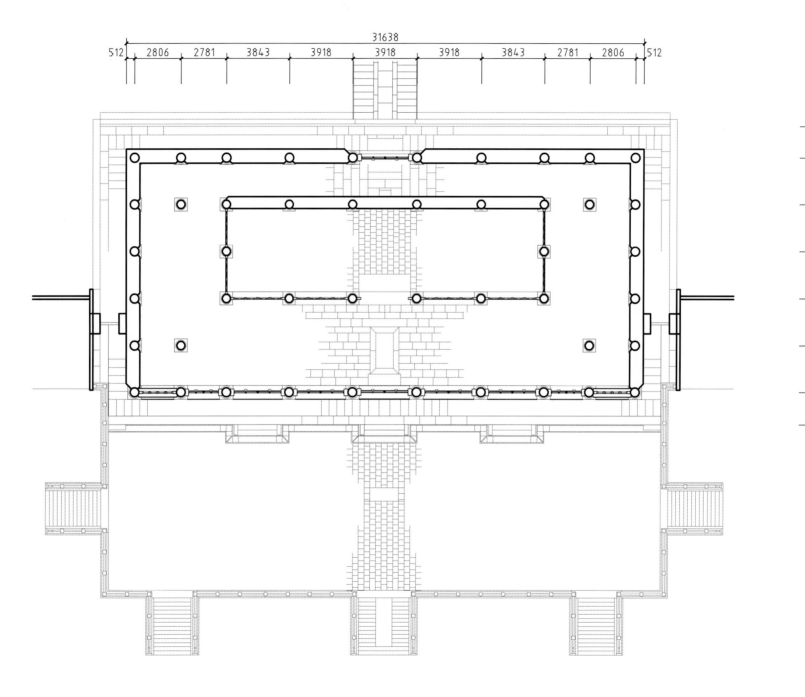

31638

512 2806 2781 3843 3918 3918 3918 3843 2781 2806 512

1850 2780 2780 17650 2780 2780 1900

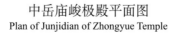

中岳庙峻极殿平面图
Plan of Junjidian of Zhongyue Temple

N

0 0.5 1m

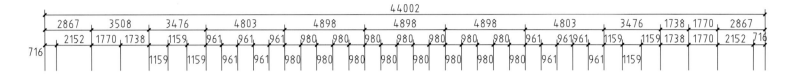

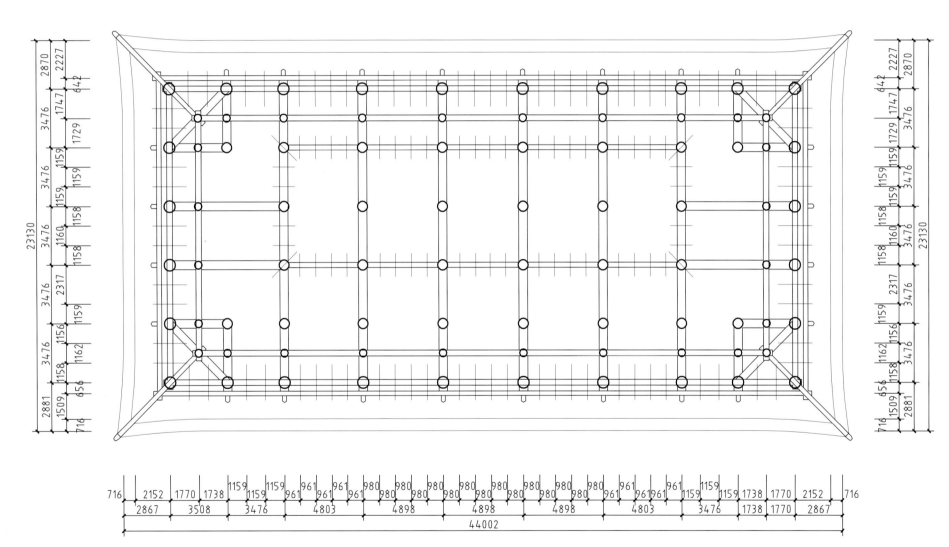

中岳庙峻极殿首层梁架俯视平面图

Plan of first-floor framework of Zhongyue Temple Junjidian as seen from above

0 1 5m

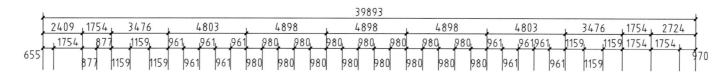

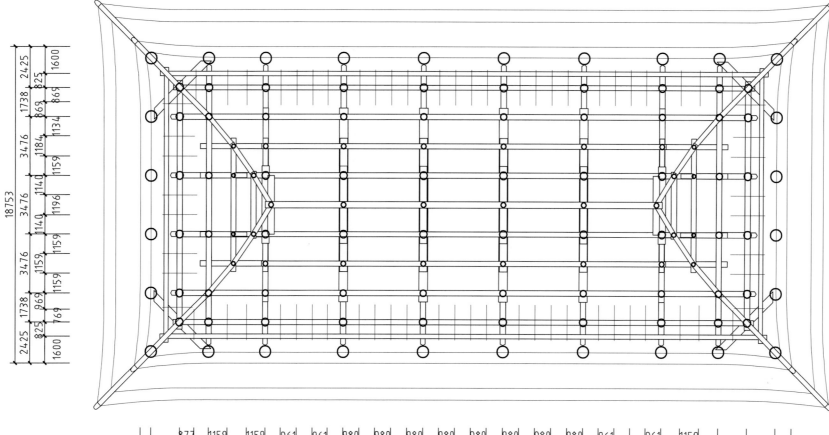

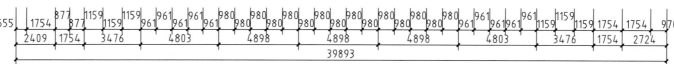

中岳庙峻极殿二层梁架俯视平面图

Plan of second-floor framework of Zhongyue Temple Junjidian as seen from above

0 1 5m

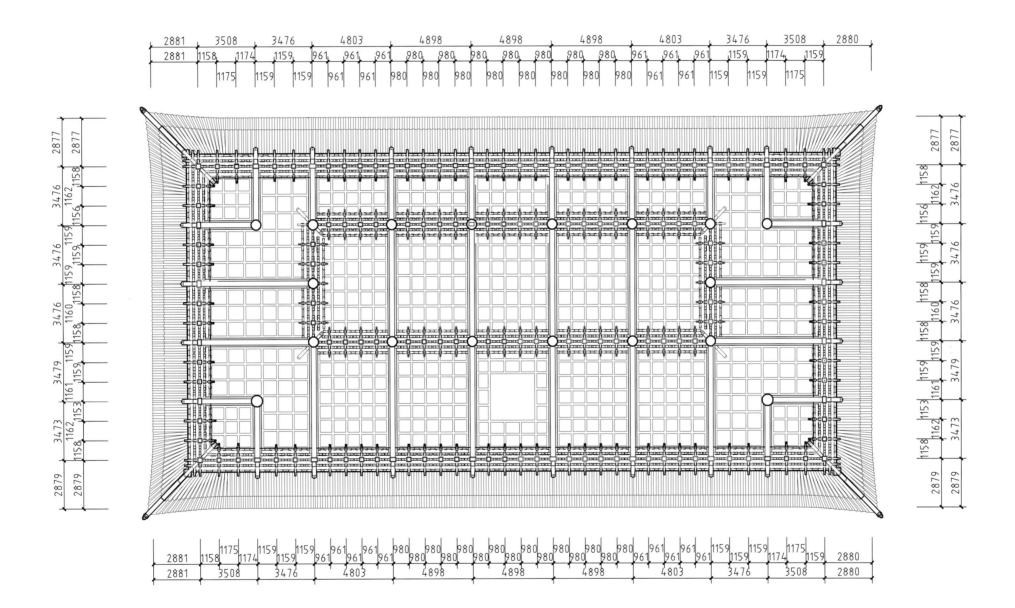

中岳庙峻极殿梁架仰视平面图

Plan of framework of Zhongyue Temple Junjidian as seen from below

0 1   5m

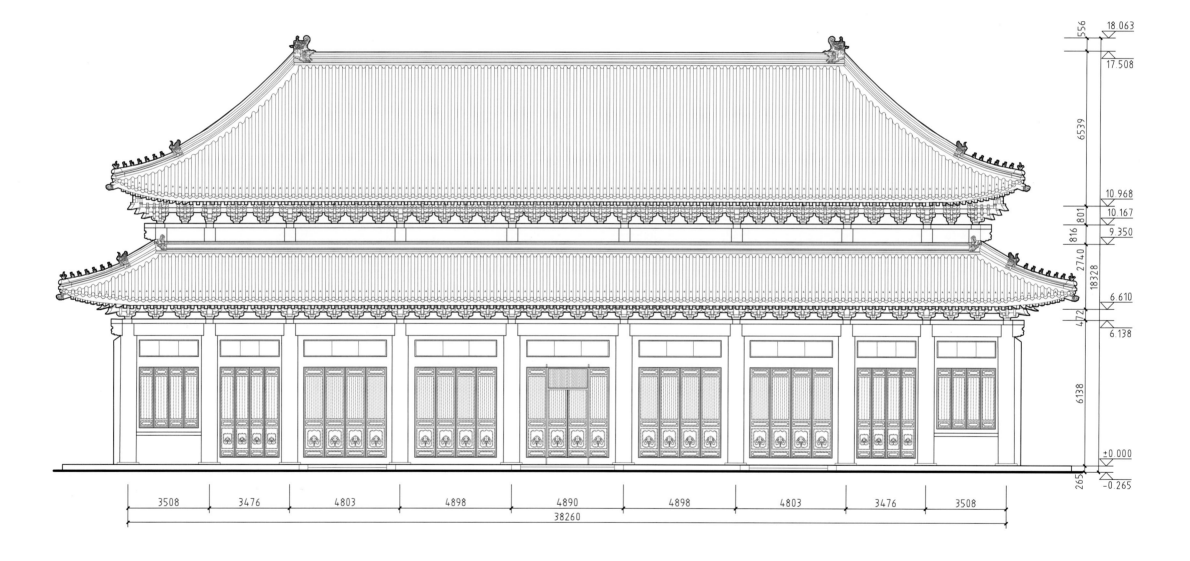

556
18 063
17 508
6539
10 968
10 167
801
816
9 350
2740
18328
6.610
472
6.138
6138
±0.000
265
-0.265

3508　3476　4803　4898　4890　4898　4803　3476　3508
38260

中岳庙峻极殿正立面图
Front elevation of Junjidian of Zhongyue Temple

0　1　　　　5m

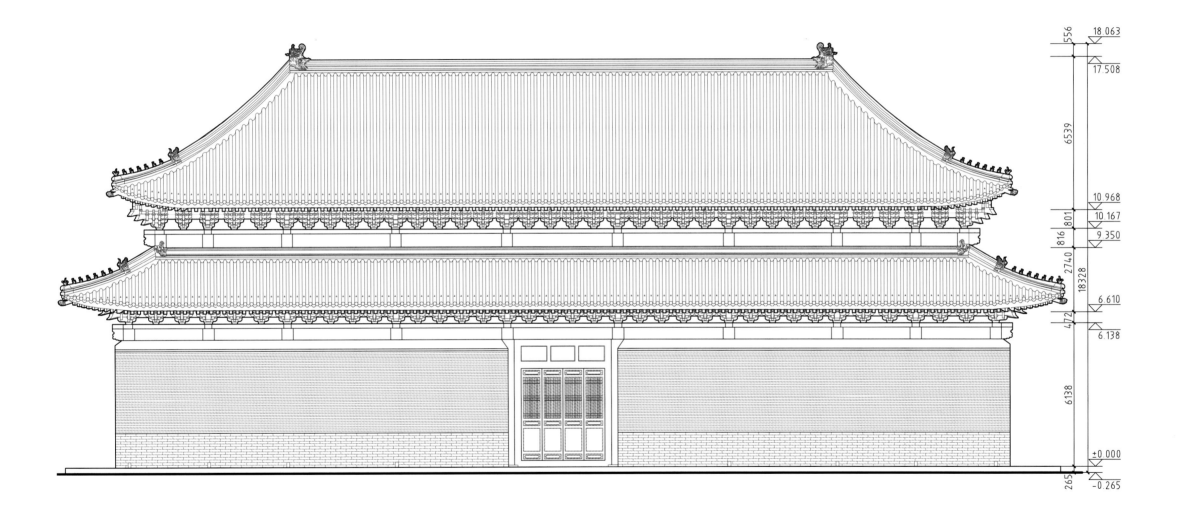

556
18.063
17.508
6539
10.968
801
10.167
816
9.350
2740
18328
6.610
472
6.138
6138
±0.000
265
-0.265

中岳庙峻极殿背立面图
Rear elevation of Junjidian of Zhongyue Temple

0 1   5m

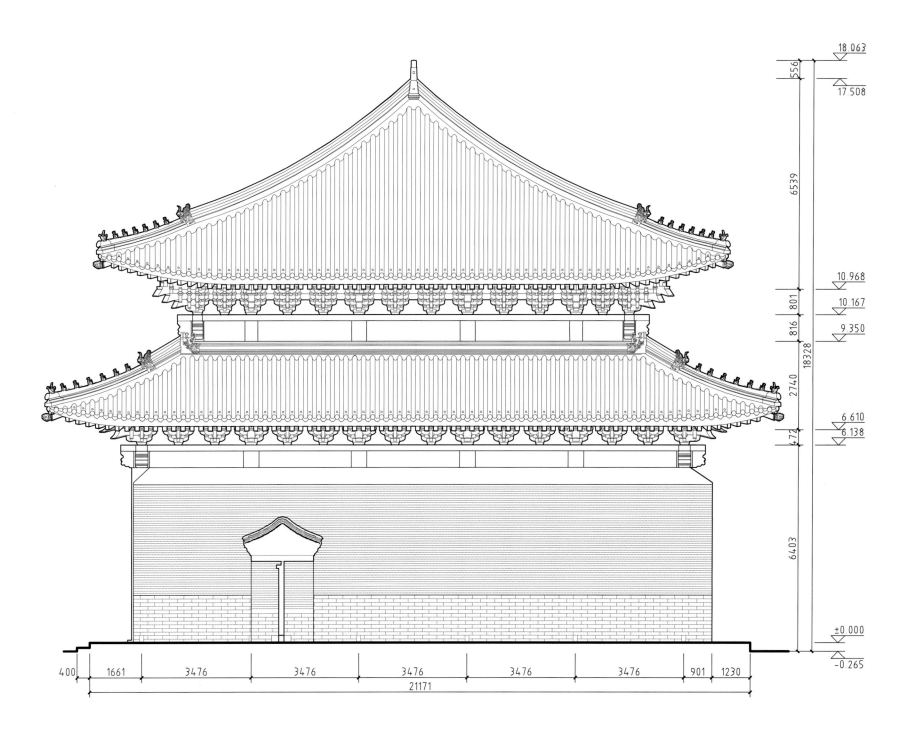

中岳庙峻极殿侧立面图
Side elevation of Junjidian of Zhongyue Temple

0 1 2m

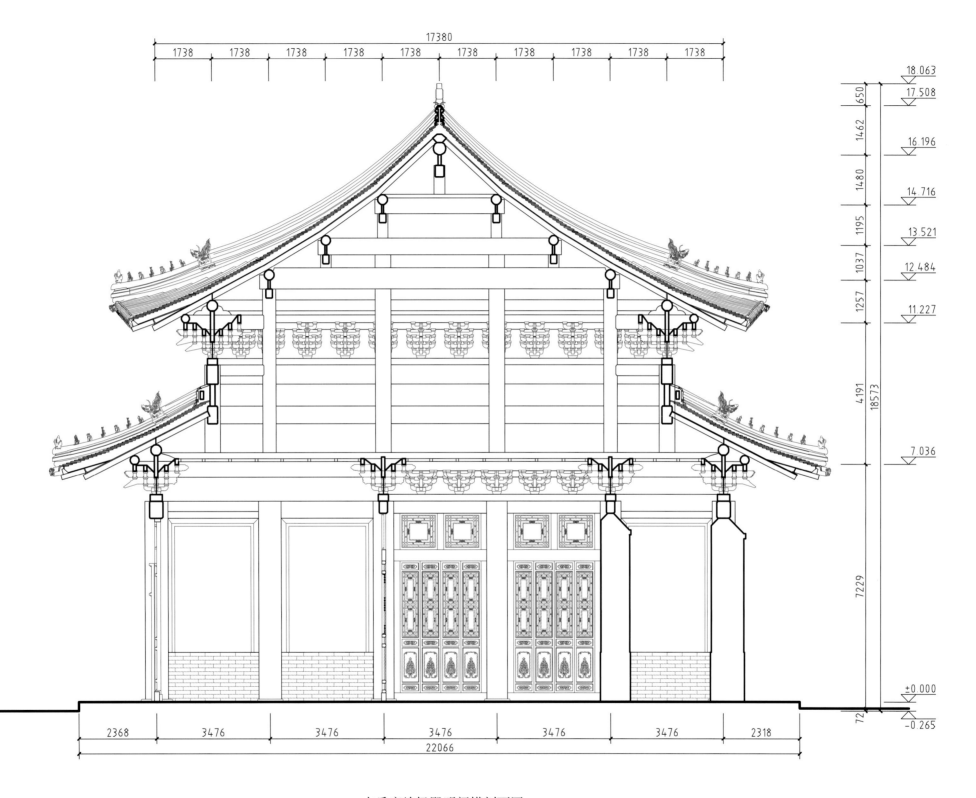

中岳庙峻极殿明间横剖面图
Cross-section of central-bay of Junjidian, Zhongyue Temple

0　1　2m

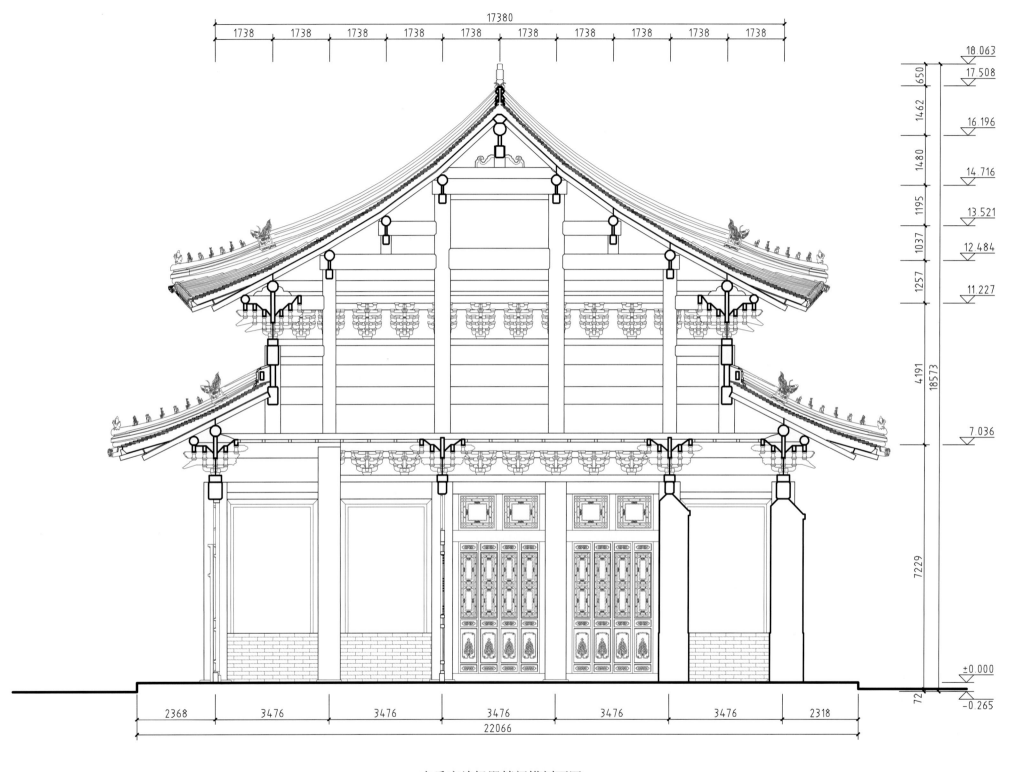

中岳庙峻极殿梢间横剖面图
Cross-section of second-to-last-bay of Junjidian, Zhongyue Temple

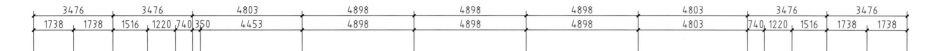

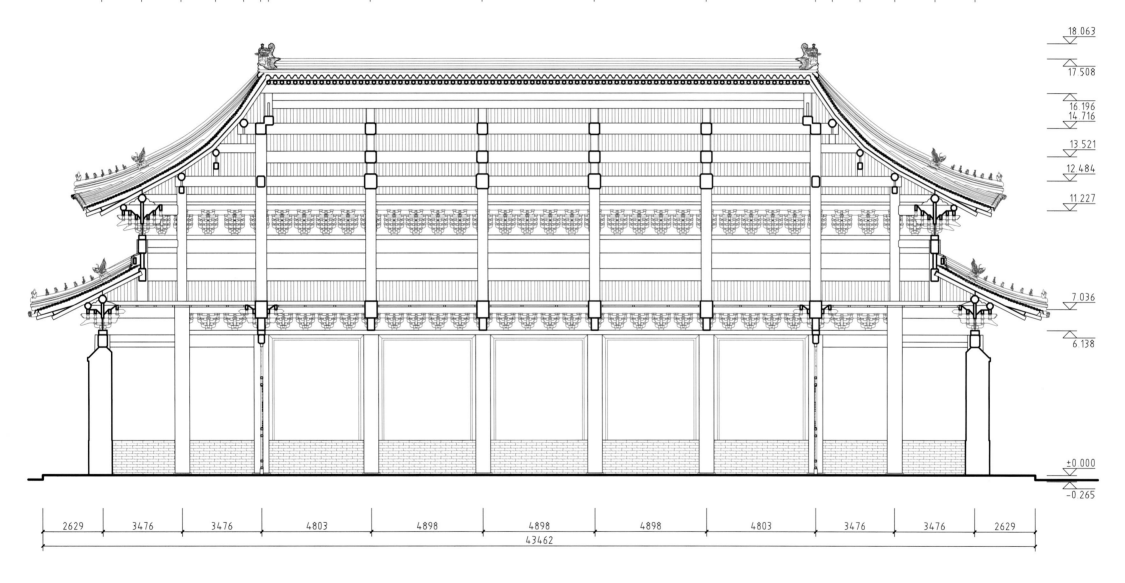

中岳庙峻极殿纵剖面图
Longitudinal section of Junjidian of Zhongyue Temple

0 1 5m

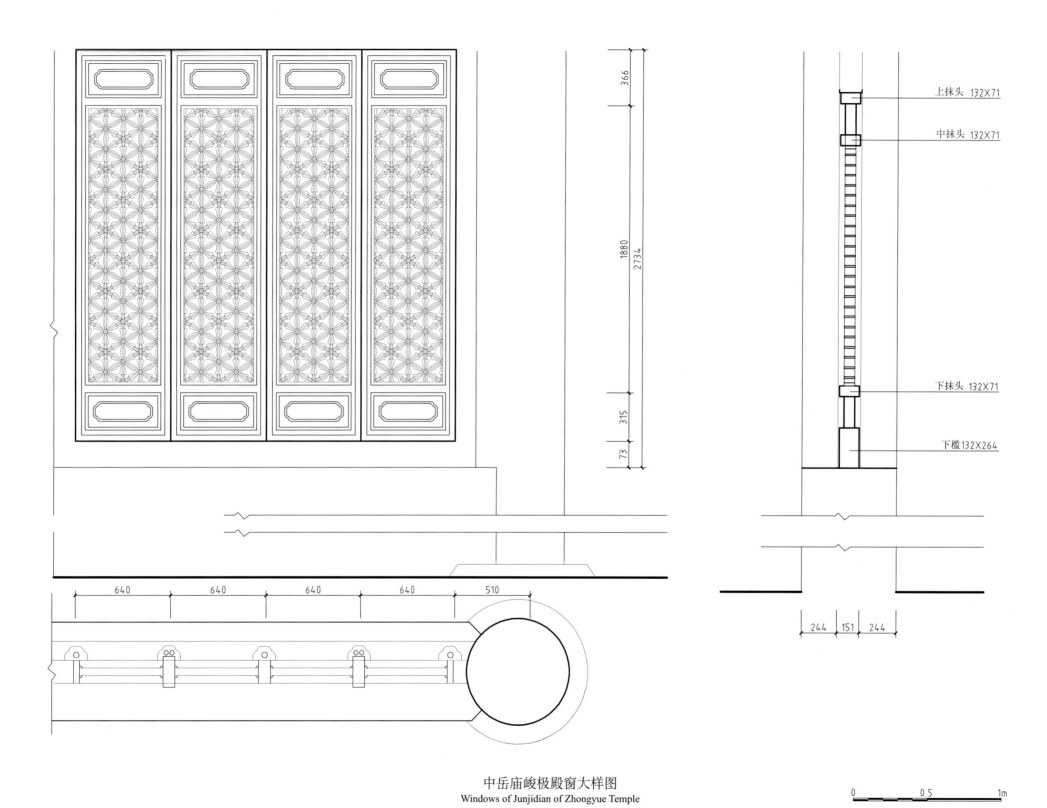

366
1880
2734
315
73

上抹头 132X71
中抹头 132X71
下抹头 132X71
下槛 132X264

640 640 640 640 510

244 151 244

中岳庙峻极殿窗大样图
Windows of Junjidian of Zhongyue Temple

0   0.5   1m

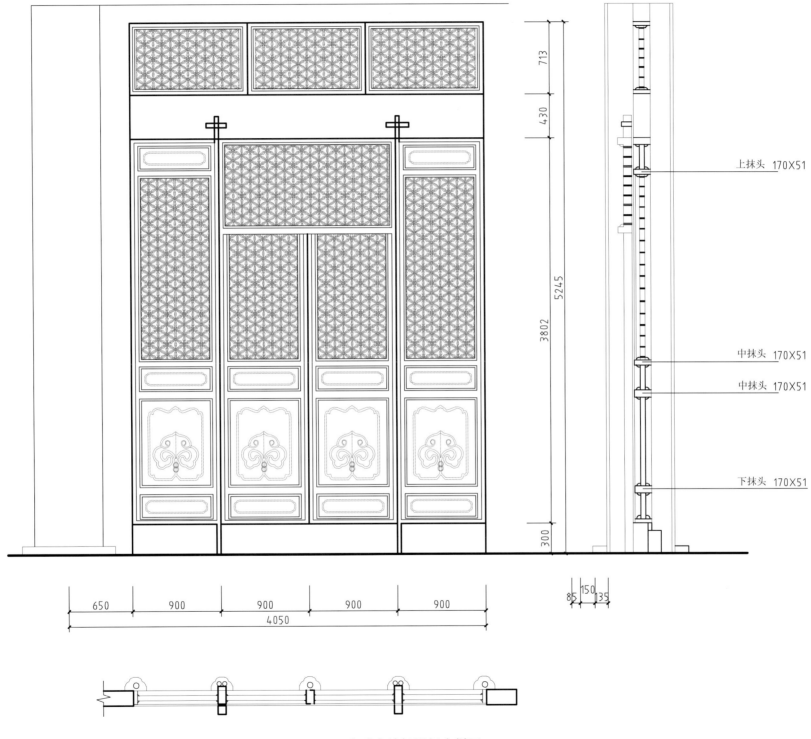

上抹头 170×51

中抹头 170×51

中抹头 170×51

下抹头 170×51

中岳庙峻极殿门大样图
Doors of Junjidian of Zhongyue Temple

0　　0.5　　1m

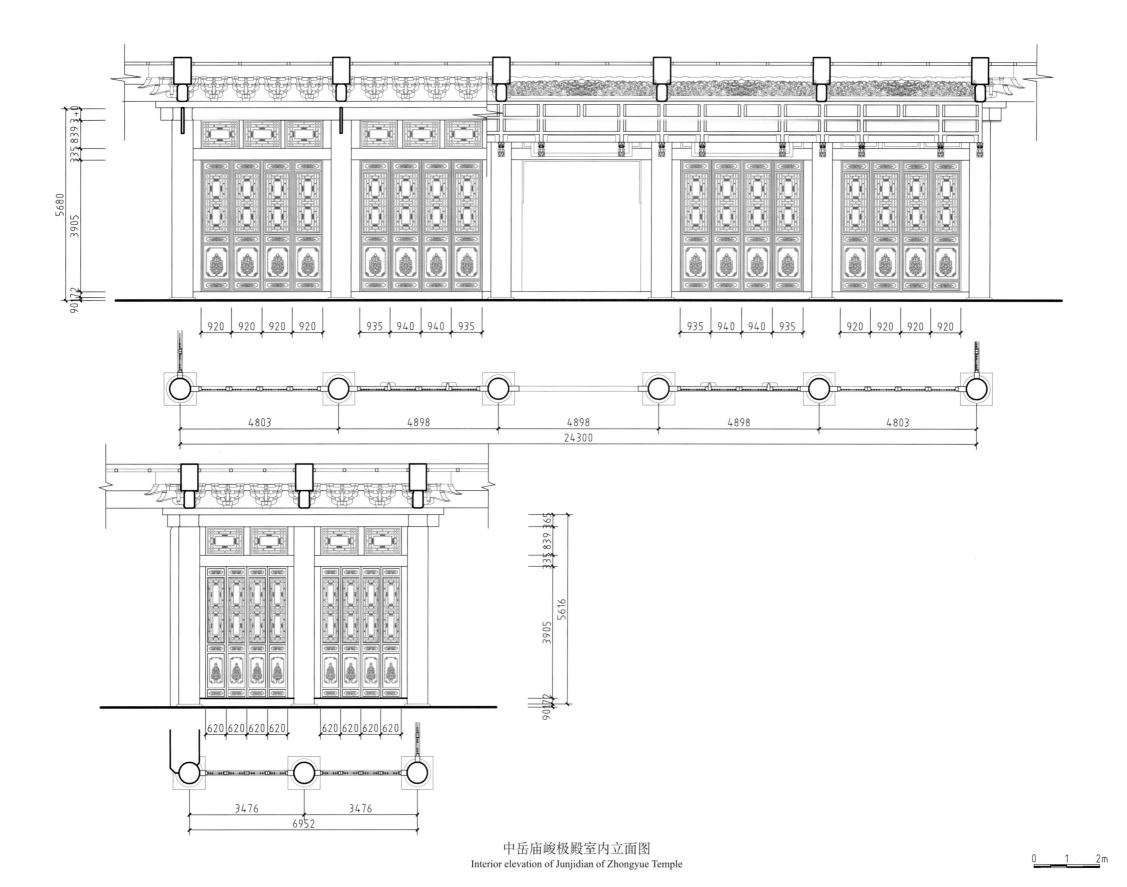

中岳庙峻极殿室内立面图
Interior elevation of Junjidian of Zhongyue Temple

0 1 2m

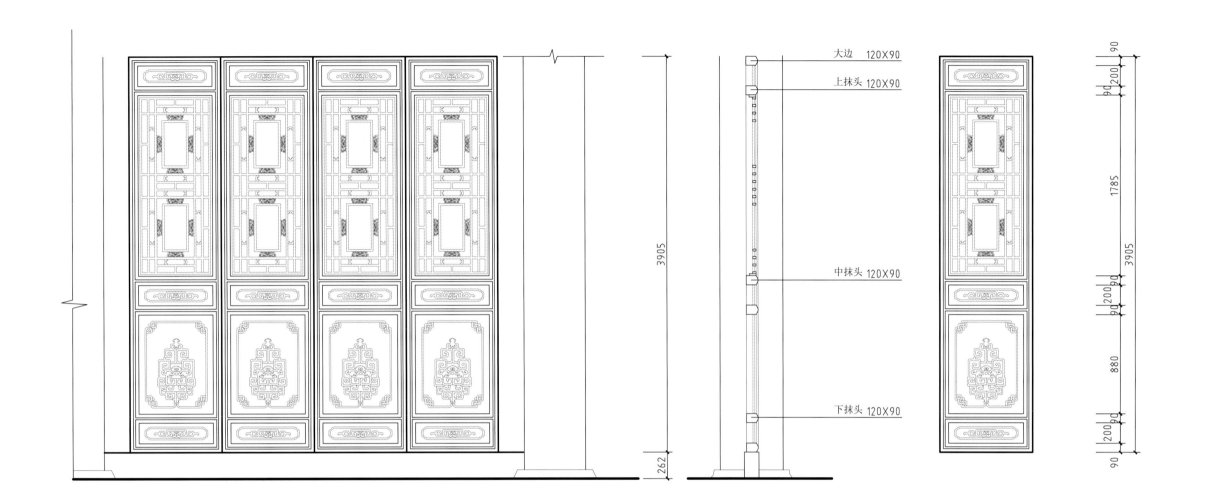

大边 120X90

上抹头 120X90

中抹头 120X90

下抹头 120X90

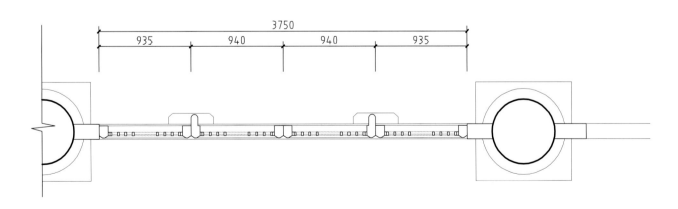

中岳庙峻极殿室内侧立面门大样图（一）
Side elevation of interior door of Junjidian, Zhongyue Temple Ⅰ

0　　　0.5　　　1m

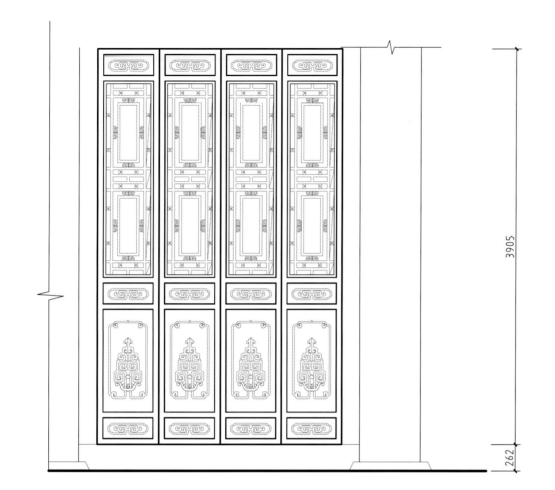

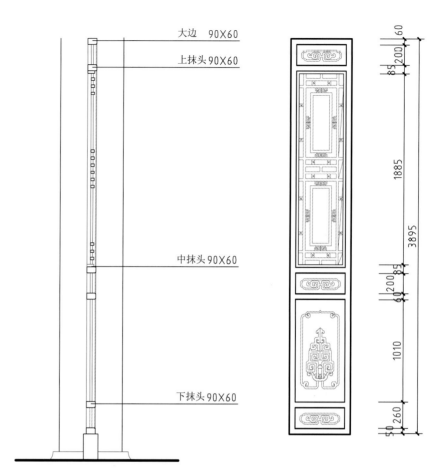

大边　90X60

上抹头90X60

中抹头90X60

下抹头90X60

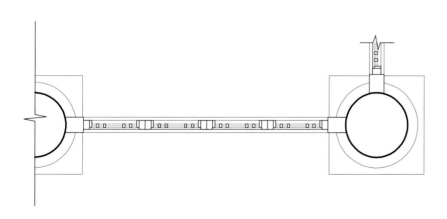

中岳庙峻极殿室内侧立面门大样图（二）
Side elevation of interior door of Junjidian, Zhongyue Temple Ⅱ

0　　0.5　　1m

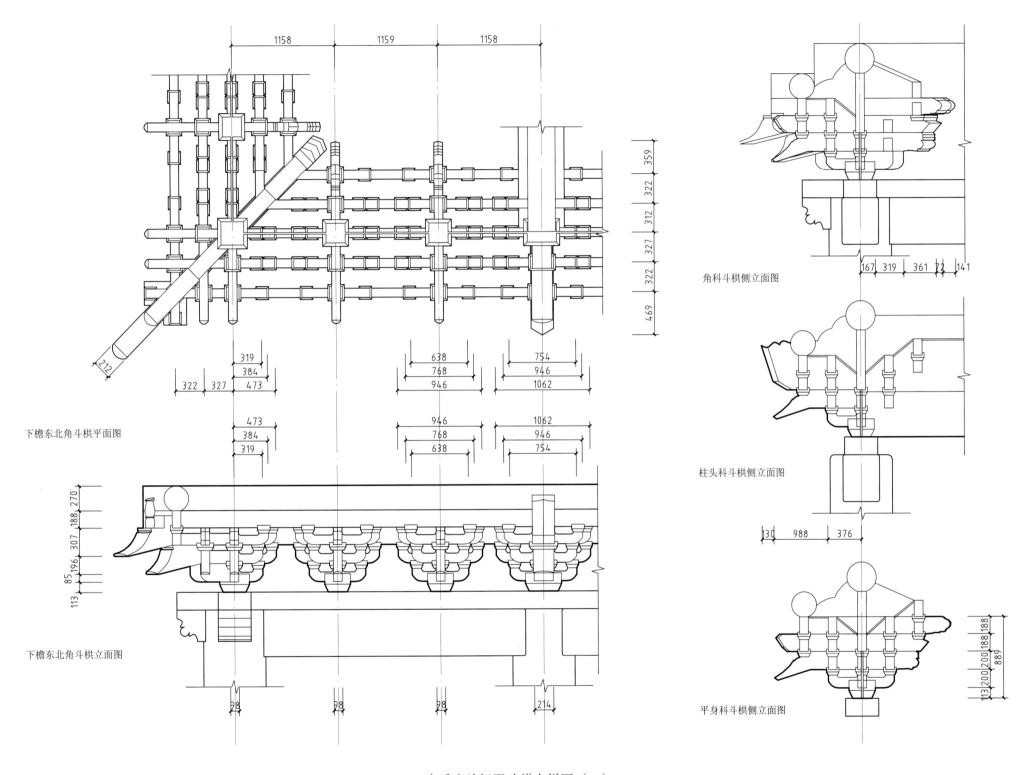

角科斗栱侧立面图

柱头科斗栱侧立面图

平身科斗栱侧立面图

下檐东北角斗栱平面图

下檐东北角斗栱立面图

中岳庙峻极殿斗栱大样图（一）

Bracket set of Junjidian of Zhongyue Temple Ⅰ

0   0.5   1m

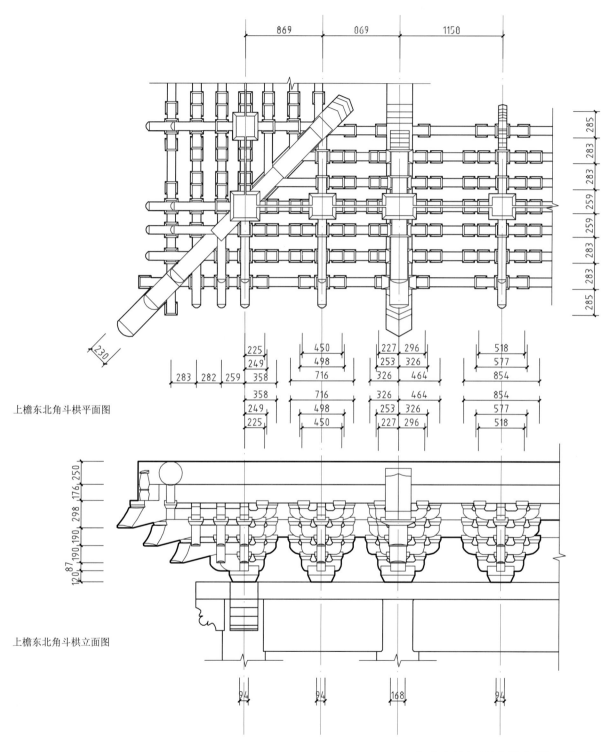

上檐东北角斗栱平面图

上檐东北角斗栱立面图

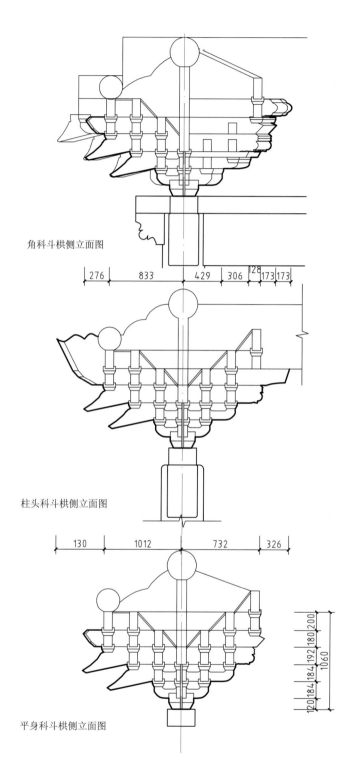

角科斗栱侧立面图

柱头科斗栱侧立面图

平身科斗栱侧立面图

中岳庙峻极殿斗栱大样图（二）

Bracket set of Junjidian of Zhongyue Temple Ⅱ

0   0.5   1m

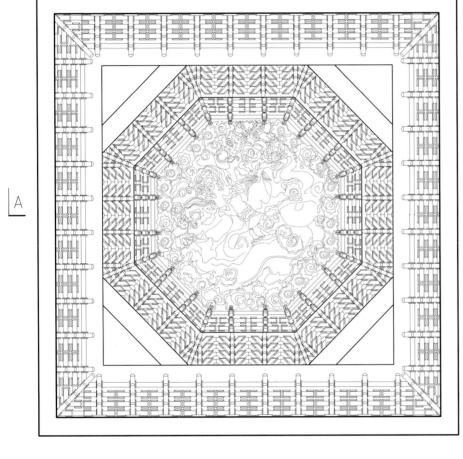

藻井仰视平面图

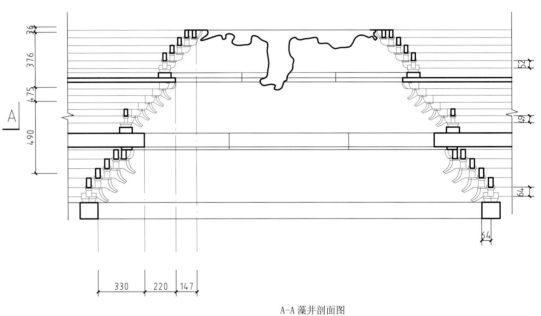

A-A 藻井剖面图

中岳庙峻极殿构件大样图
Structural components of Junjidian of Zhongyue Temple

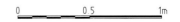

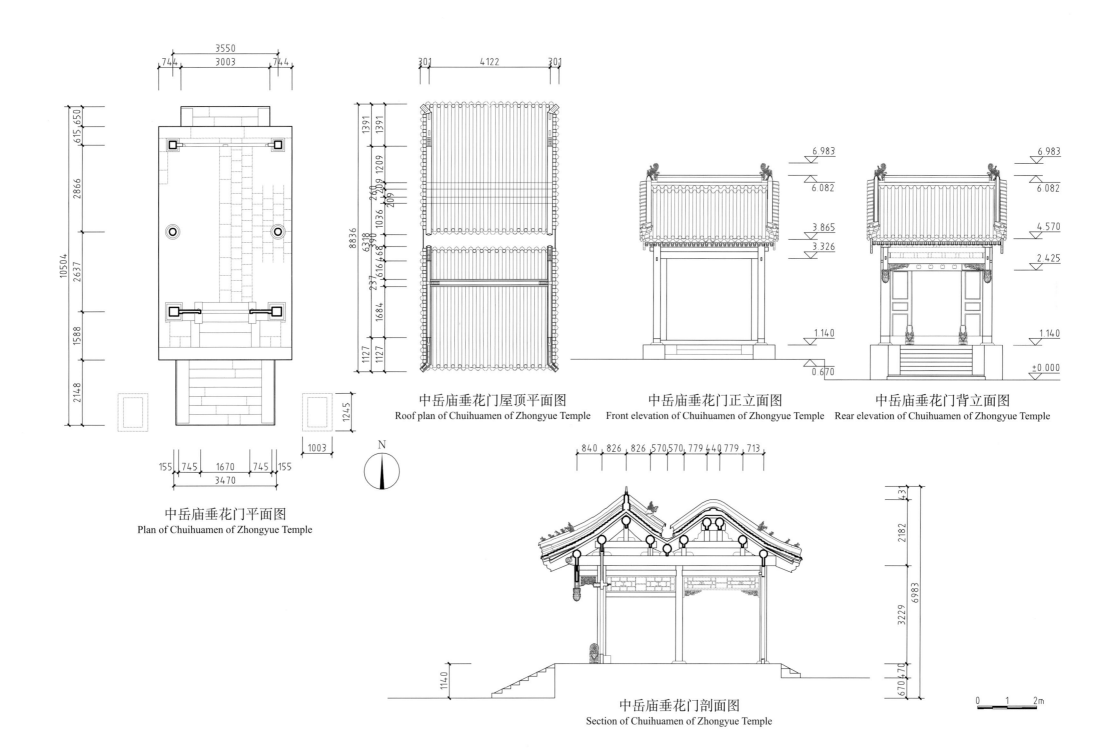

中岳庙垂花门平面图
Plan of Chuihuamen of Zhongyue Temple

中岳庙垂花门屋顶平面图
Roof plan of Chuihuamen of Zhongyue Temple

中岳庙垂花门正立面图
Front elevation of Chuihuamen of Zhongyue Temple

中岳庙垂花门背立面图
Rear elevation of Chuihuamen of Zhongyue Temple

中岳庙垂花门剖面图
Section of Chuihuamen of Zhongyue Temple

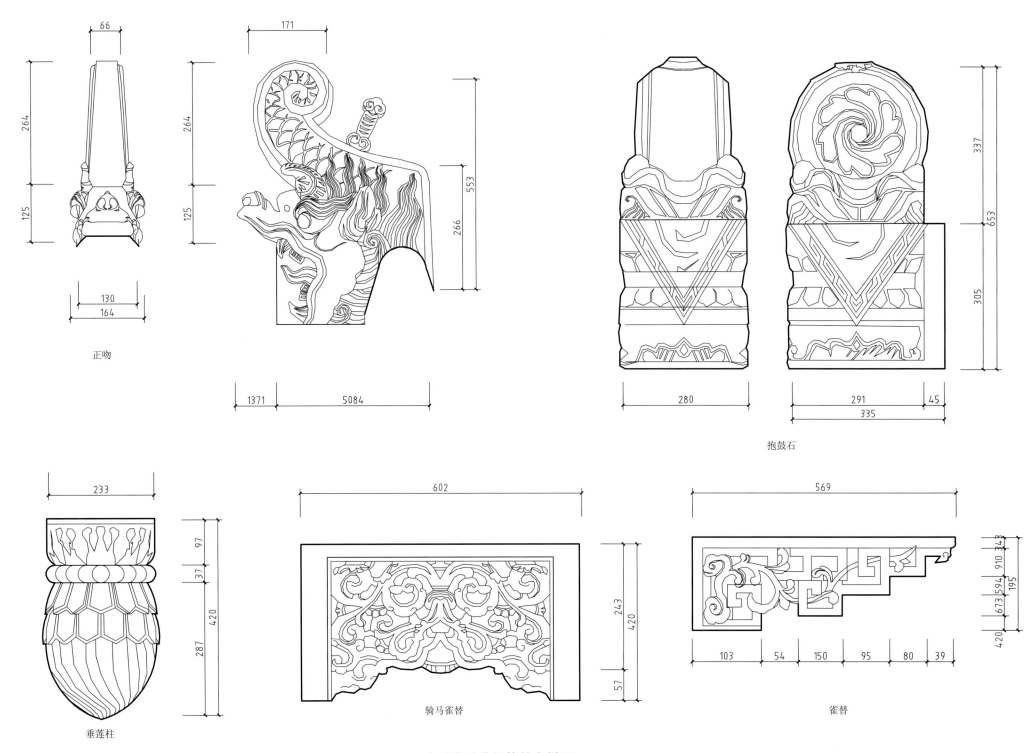

正吻

抱鼓石

垂莲柱

骑马雀替

雀替

中岳庙垂花门构件大样图
Structural components of Chuihuamen of Zhongyue Temple

0   1   2m

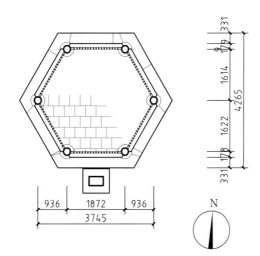

中岳庙柏亭平面图
Plan of Baiting of Zhongyue Temple

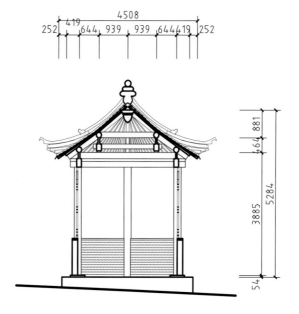

中岳庙柏亭剖面图
Section of Baiting of Zhongyue Temple

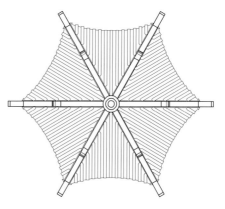

中岳庙柏亭屋顶平面图
Roof plan of Baiting of Zhongyue Temple

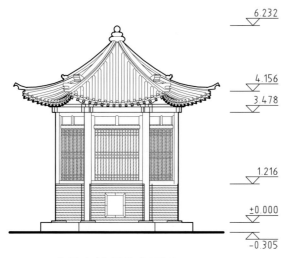

中岳庙柏亭北立面图
North elevation of Baiting of Zhongyue Temple

中岳庙柏亭西立面图
West elevation of Baiting of Zhongyue Temple

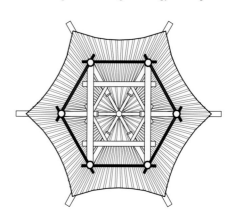

中岳庙柏亭梁架仰视平面图
Framework plan of Baiting of Zhongyue Temple as seen from below

0   1   2m

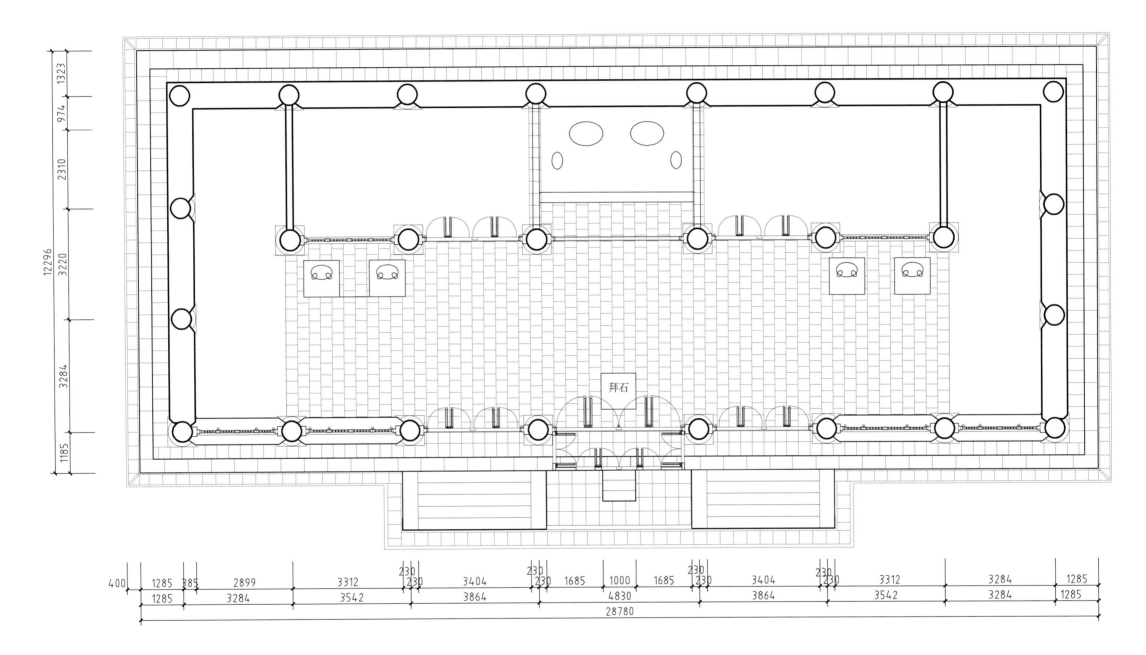

拜石

中岳庙寝殿平面图

Plan of *qindian* of Zhongyue Temple

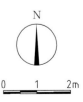

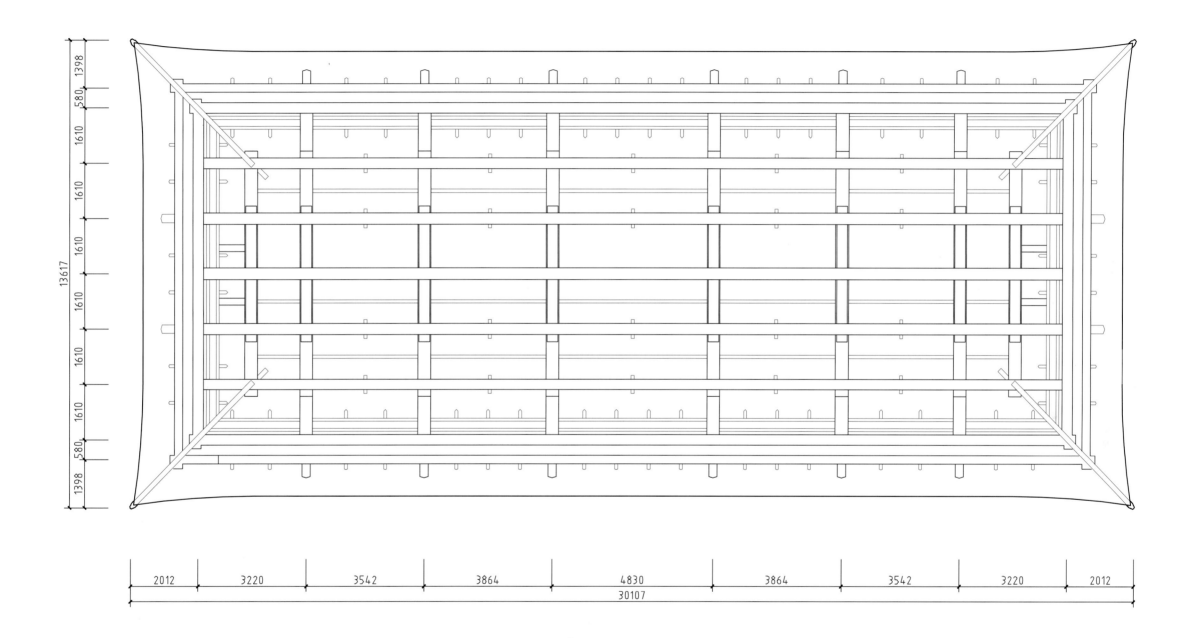

中岳庙寝殿梁架俯视平面图
Framework plan of *qindian* of Zhongyue Temple as seen from above

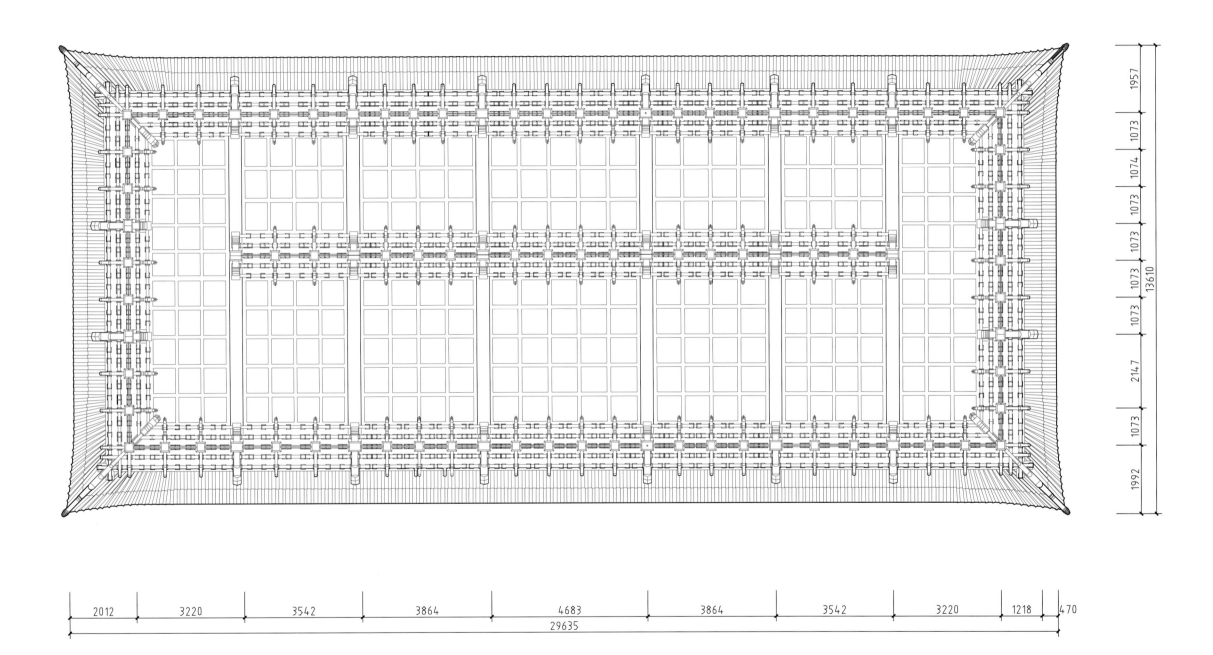

中岳庙寝殿梁架仰视平面图

Framework plan of *qindian* of Zhongyue Temple as seen from below

0　1　2m

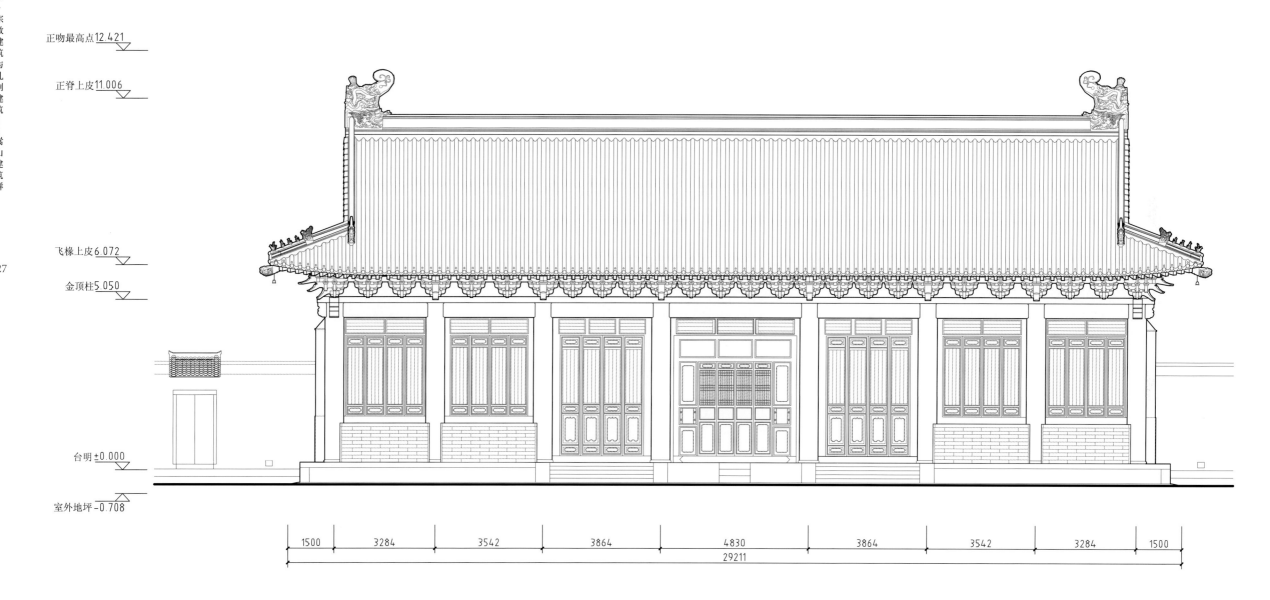

正吻最高点12.421

正脊上皮11.006

飞椽上皮6.072

金顶柱5.050

台明±0.000

室外地坪-0.708

1500　3284　3542　3864　4830　3864　3542　3284　1500

29211

中岳庙寝殿正立面图

Front elevation of *qindian* of Zhongyue Temple

0　1　2m

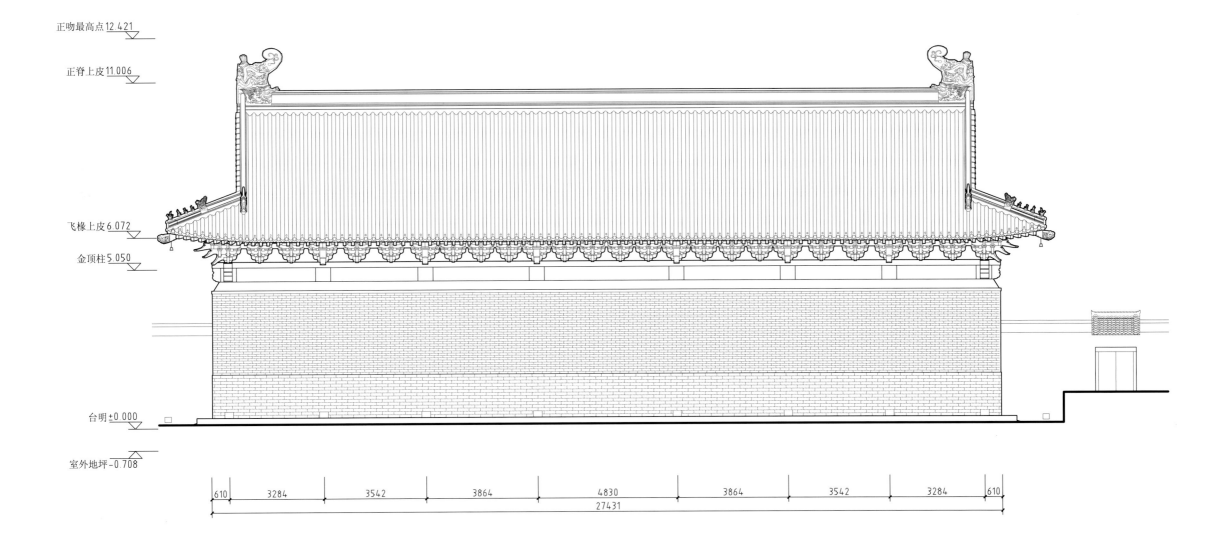

正吻最高点12.421

正脊上皮11.006

飞椽上皮6.072

金顶柱5.050

台明±0.000

室外地坪-0.708

| 610 | 3284 | 3542 | 3864 | 4830 | 3864 | 3542 | 3284 | 610 |

27431

中岳庙寝殿背立面图
Rear elevation of *qindian* of Zhongyue Temple

0　　1　　2m

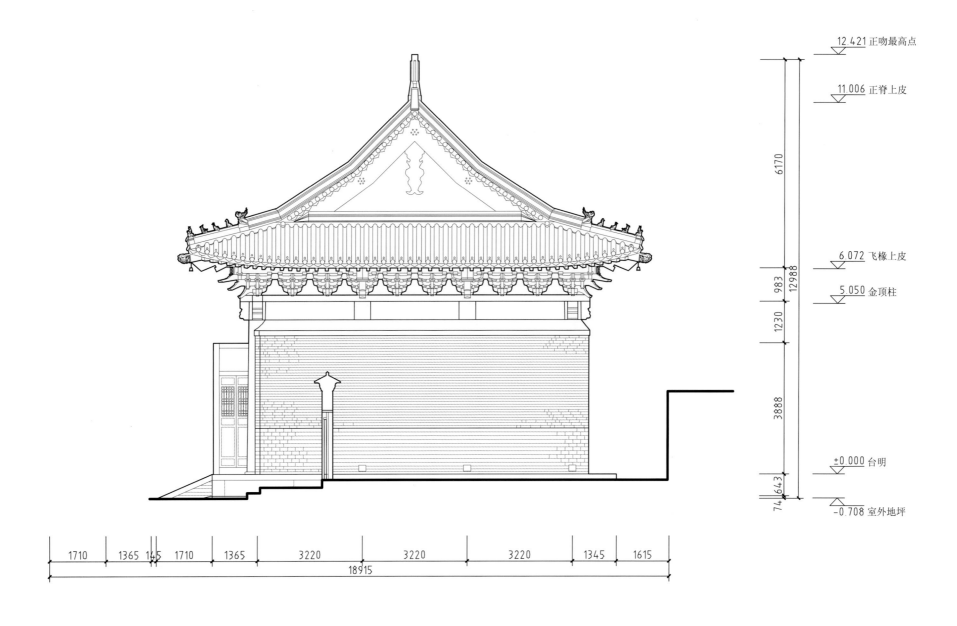

12.421 正吻最高点

11.006 正脊上皮

6.072 飞椽上皮

5.050 金顶柱

±0.000 台明

-0.708 室外地坪

6170

983

1230

12988

3888

74 643

1710 1365 145 1710 1365 3220 3220 3220 1345 1615

18915

中岳庙寝殿侧立面图
Side elevation of *qindian* of Zhongyue Temple

0   1   2m

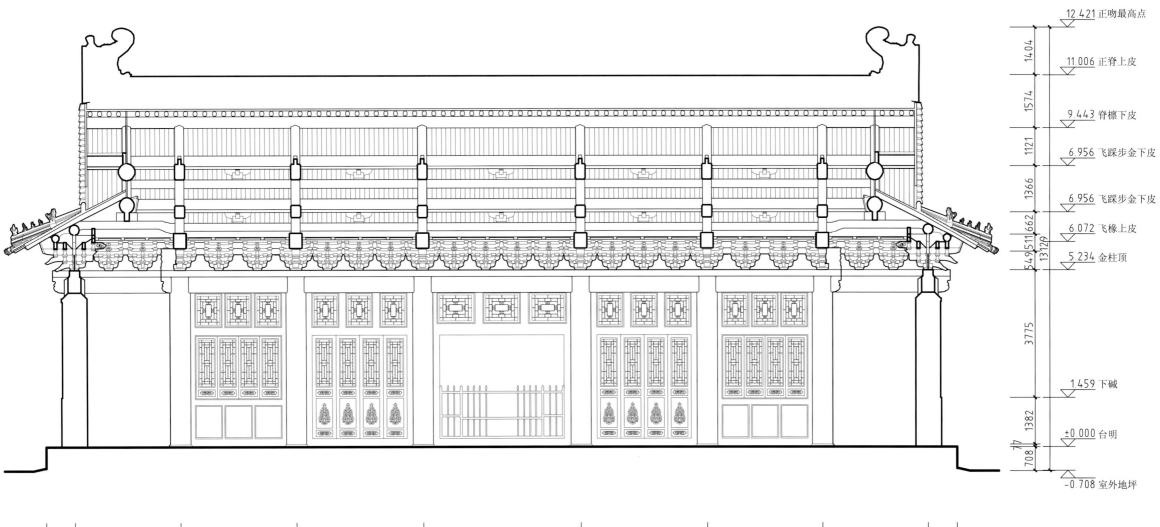

345 361 865 580 337

25408

580 865 361 345

12.421 正吻最高点

1404

11.006 正脊上皮

1574

9.443 脊檩下皮

1121

6.956 飞踩步金下皮

1366

6.956 飞踩步金下皮

662

6.072 飞椽上皮

549 511 1312 9

5.234 金柱顶

3775

1.459 下碱

1382

±0.000 台明

77

708

-0.708 室外地坪

230

880 3220 3542 3864 4830 3864 3542 3220 880

27842

中岳庙寝殿纵剖面图
Longitudinal section of *qindian* of Zhongyue Temple

0 1 2m

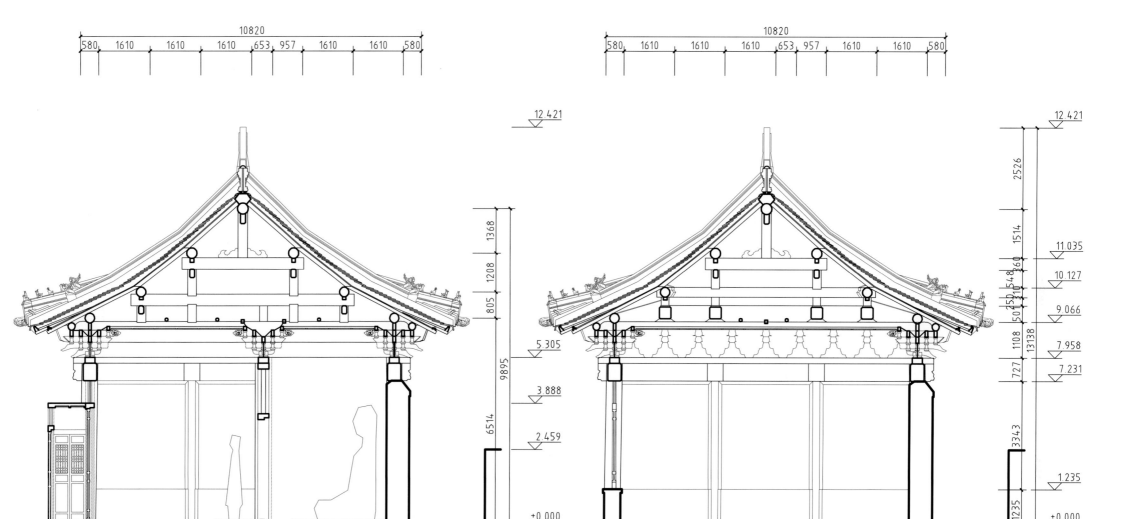

中岳庙寝殿明间横剖面图
Cross-section of central-bay of *qindian*, Zhongyue Temple

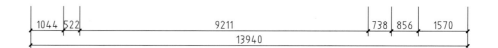

中岳庙寝殿梢间横剖面图
Cross-section of second-to-last-bay of *qindian*, Zhongyue Temple

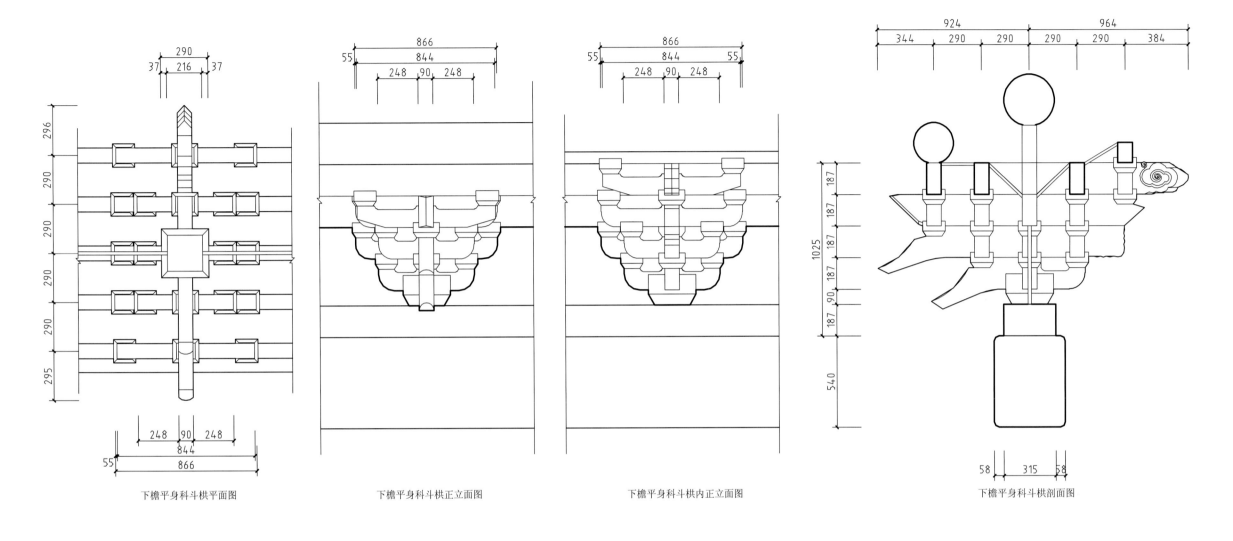

下檐平身科斗栱平面图　　　　　下檐平身科斗栱正立面图　　　　　下檐平身科斗栱内正立面图　　　　　下檐平身科斗栱剖面图

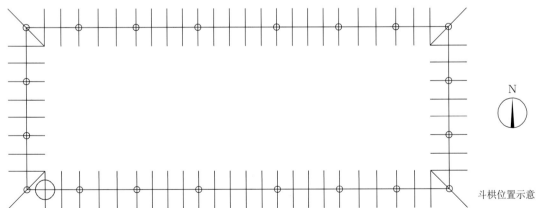

斗栱位置示意图

N

中岳庙寝殿斗栱大样图（一）

Bracket set of *qindian* of Zhongyue Temple　Ⅰ

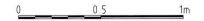

0　　　　　0.5　　　　　1m

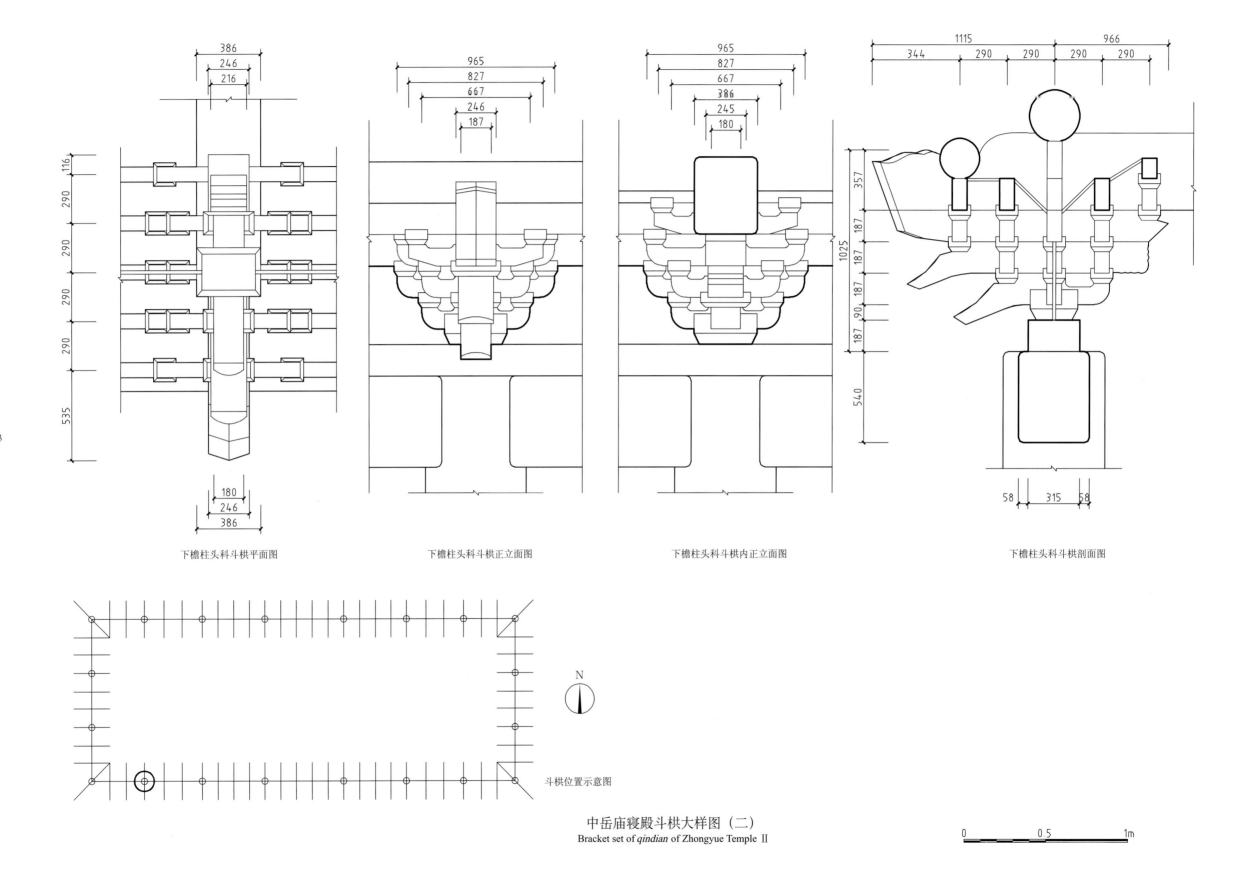

下檐柱头科斗栱平面图

下檐柱头科斗栱正立面图

下檐柱头科斗栱内正立面图

下檐柱头科斗栱剖面图

斗栱位置示意图

中岳庙寝殿斗栱大样图（二）
Bracket set of *qindian* of Zhongyue Temple Ⅱ

0   0.5   1m

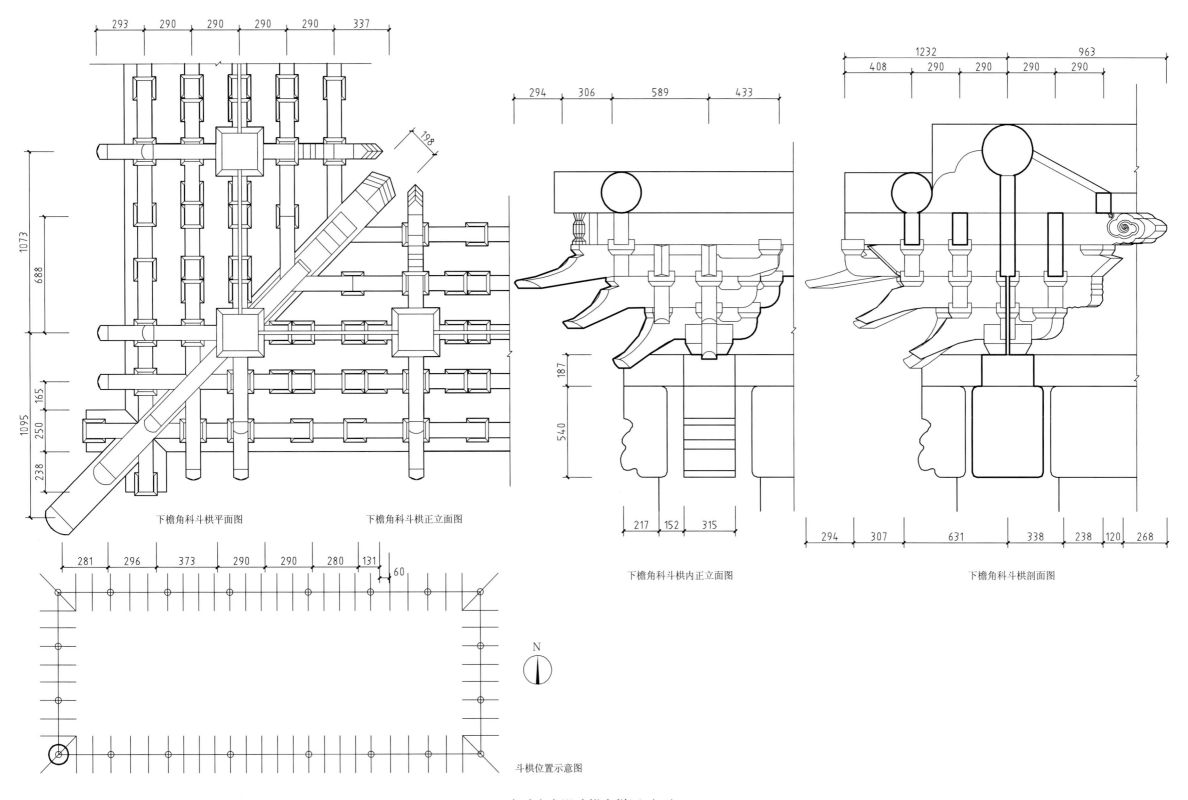

下檐角科斗栱平面图

下檐角科斗栱正立面图

下檐角科斗栱内正立面图

下檐角科斗栱剖面图

斗栱位置示意图

N

中岳庙寝殿斗栱大样图（三）
Bracket set of *qindian* of Zhongyue Temple Ⅲ

0　　　　0.5　　　　1m

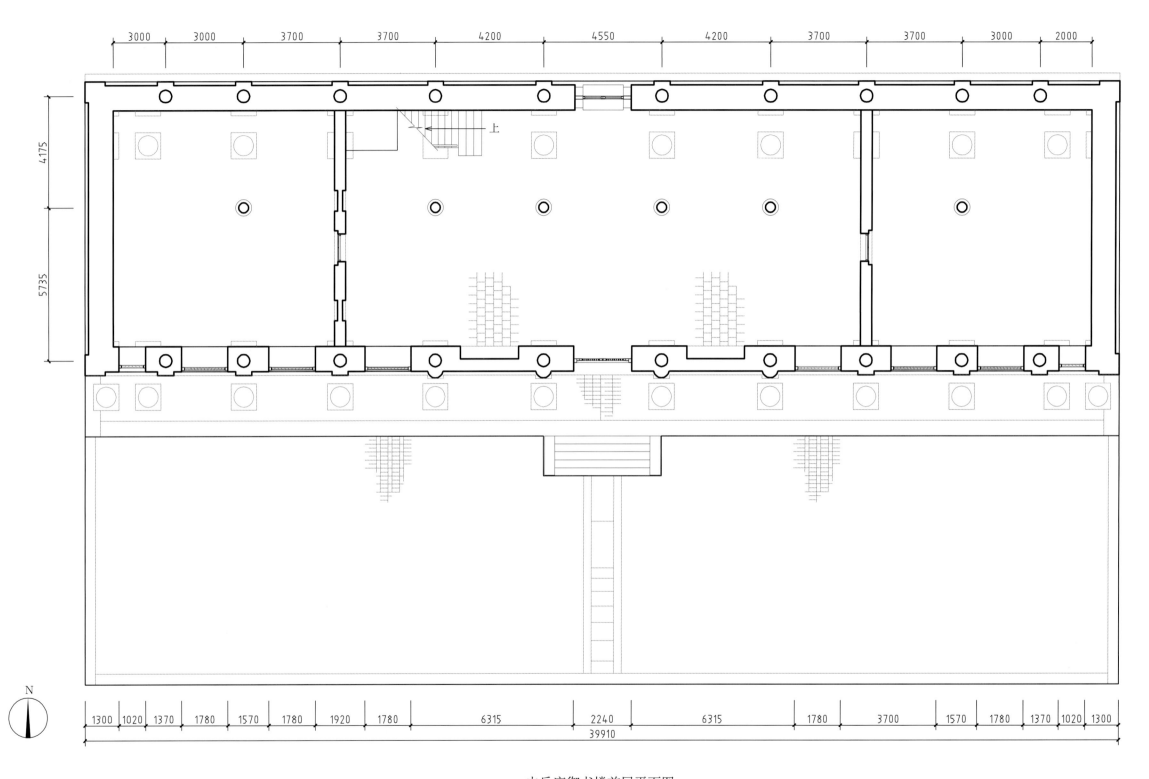

中岳庙御书楼首层平面图
Plan of first floor of Yushulou, Zhongyue Temple

N

0 1 2m

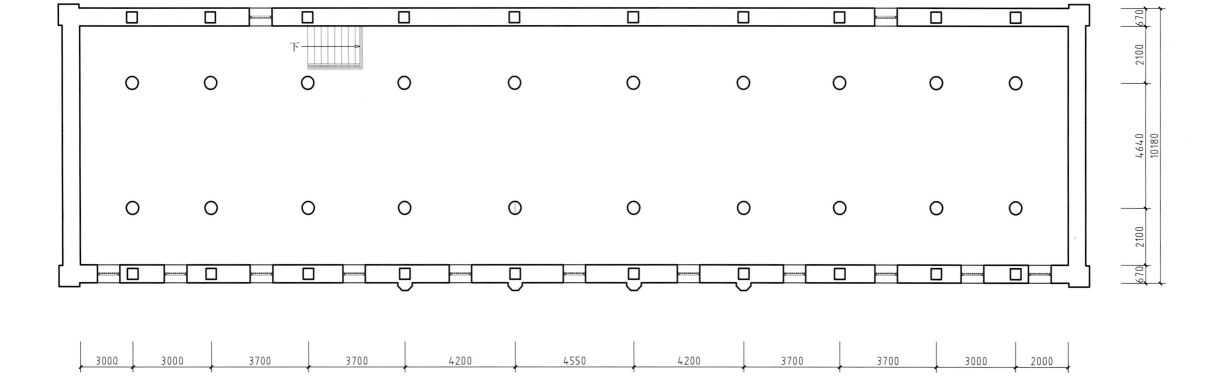

下

670
2100
4640
10180
2100
670

3000　3000　3700　3700　4200　4550　4200　3700　3700　3000　2000

中岳庙御书楼二层平面图
Plan of second floor of Yushulou, Zhongyue Temple

0　1　2m

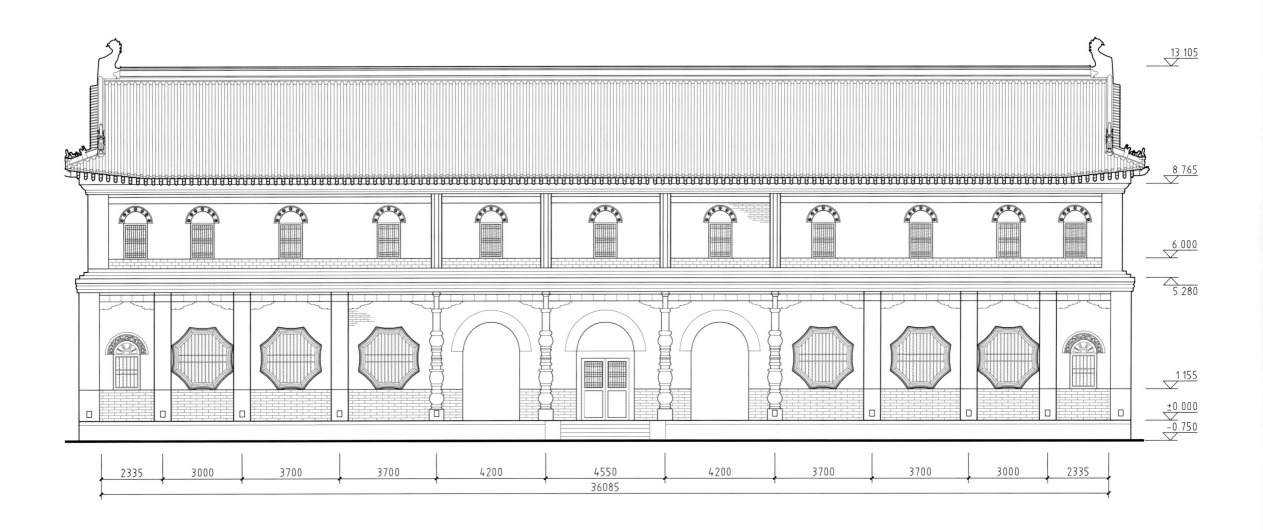

13.105

8.765

6.000

5.280

1.155

±0.000

-0.750

2335　3000　3700　3700　4200　4550　4200　3700　3700　3000　2335

36085

中岳庙御书楼正立面图

Front elevation of Yushulou of Zhongyue Temple

0　1　2m

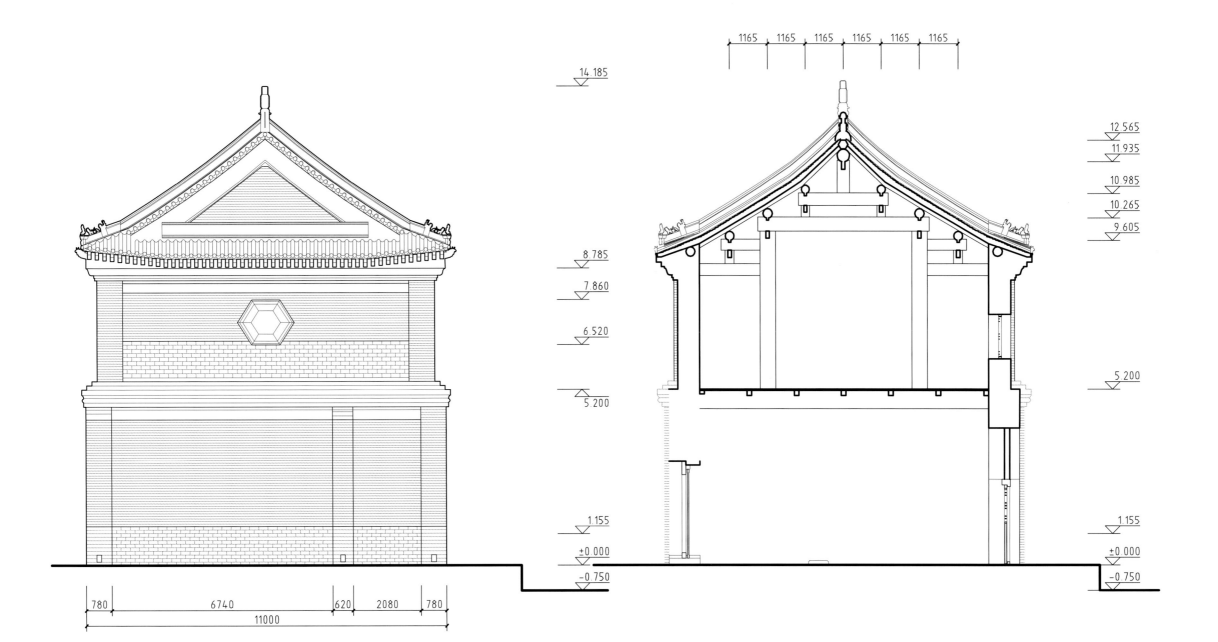

14.185

12.565
11.935
10.985
10.265
9.605

1165 1165 1165 1165 1165 1165

8.785
7.860

6.520

5.200

5.200

1.155

1.155

±0.000

±0.000

-0.750

-0.750

780 6740 620 2080 780

11000

中岳庙御书楼侧立面图
Side elevation of Yushulou of Zhongyue Temple

中岳庙御书楼横剖面图
Cross-section of Yushulou of Zhongyue Temple

0 1 2m

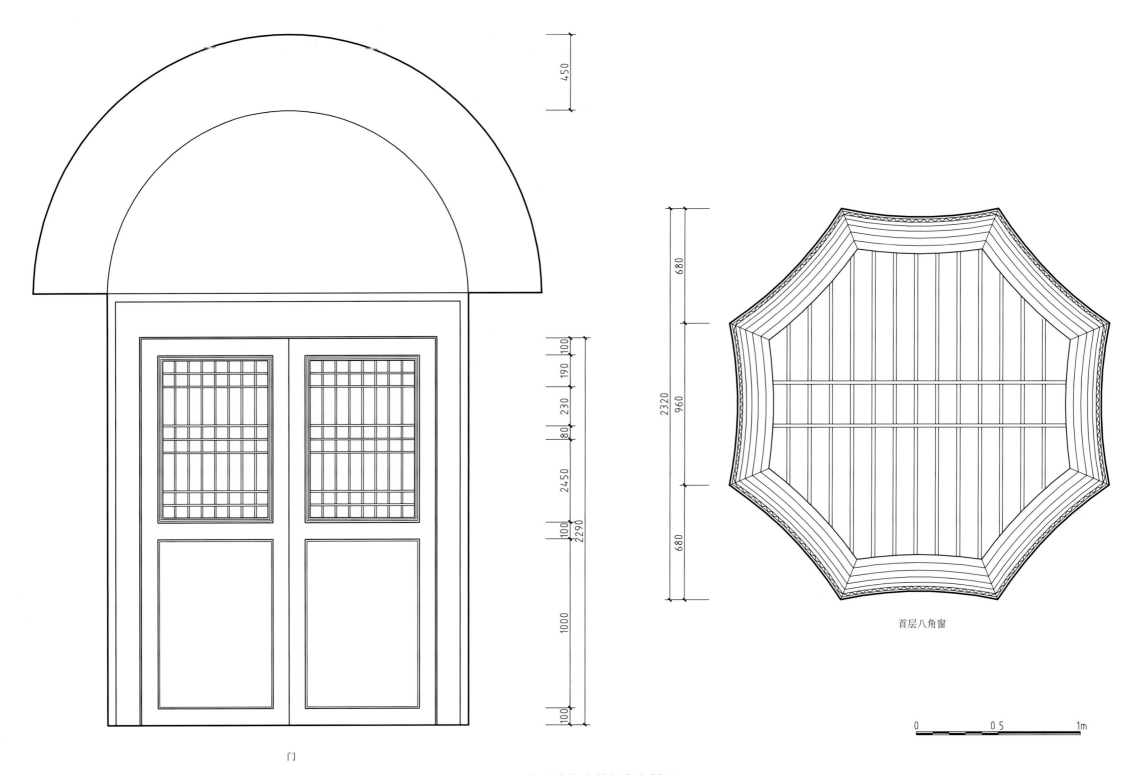

门

首层八角窗

中岳庙御书楼门窗大样图
Doors and windows of Yushulou of Zhongyue Temple

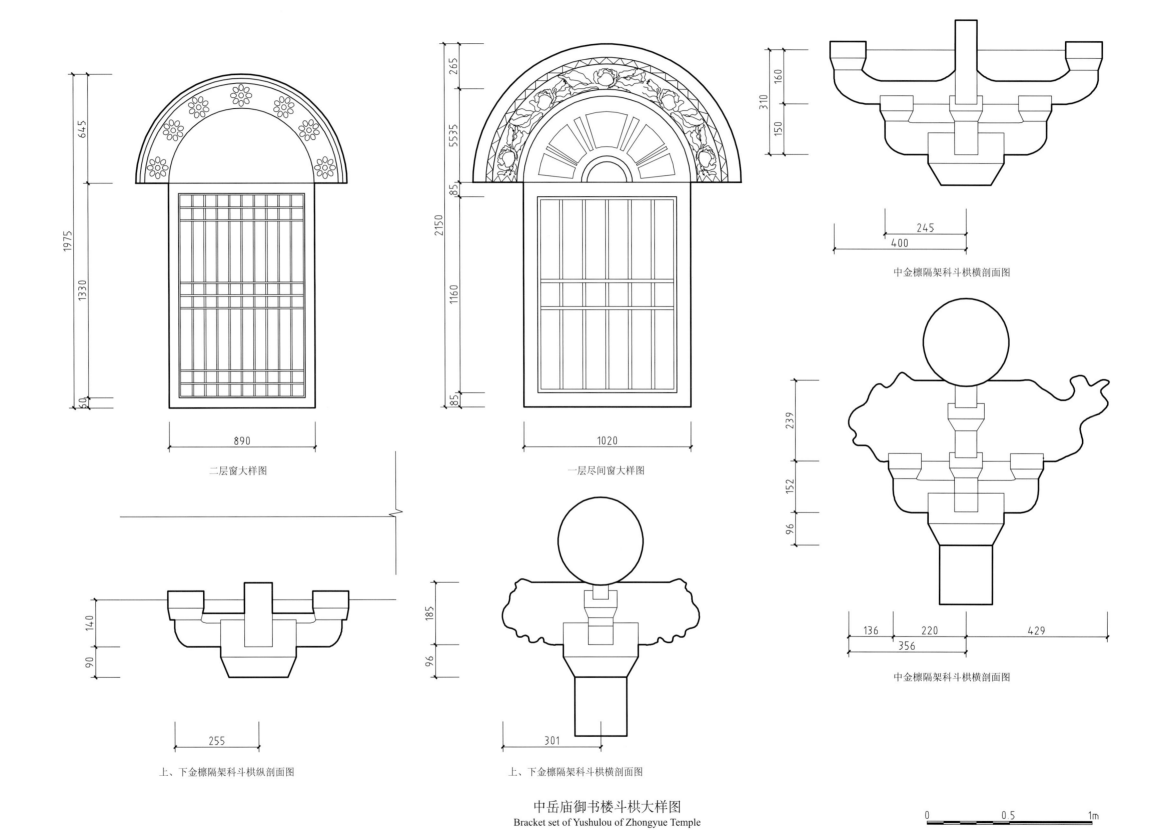

中金檩隔架科斗栱横剖面图

二层窗大样图

一层尽间窗大样图

中金檩隔架科斗栱横剖面图

上、下金檩隔架科斗栱纵剖面图

上、下金檩隔架科斗栱横剖面图

中岳庙御书楼斗栱大样图
Bracket set of Yushulou of Zhongyue Temple

0　　　　　0.5　　　　　1m

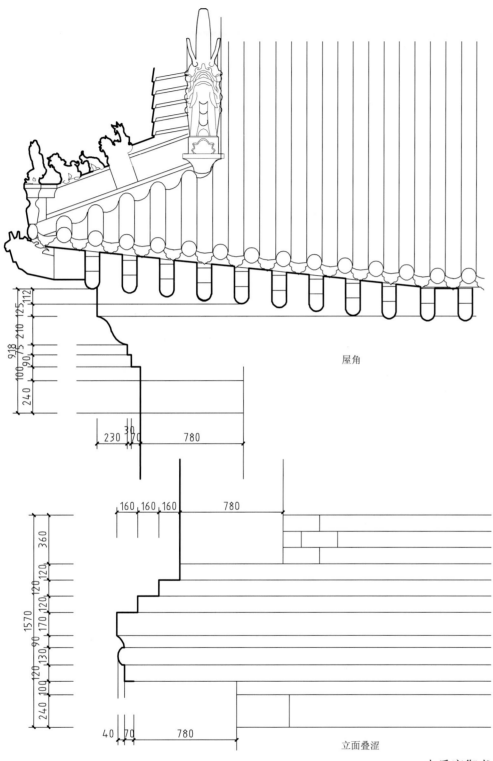

屋角

立面叠涩

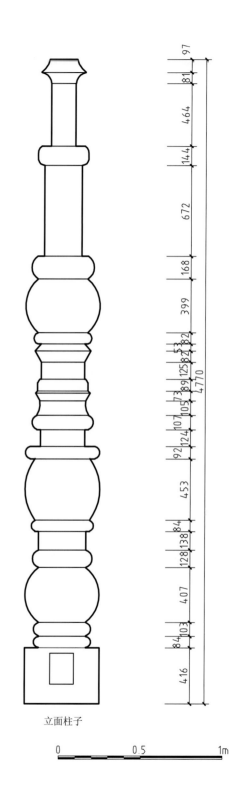

立面柱子

中岳庙御书楼构件大样图（一）
Structural components of Yushulou of Zhongyue Temple Ⅰ

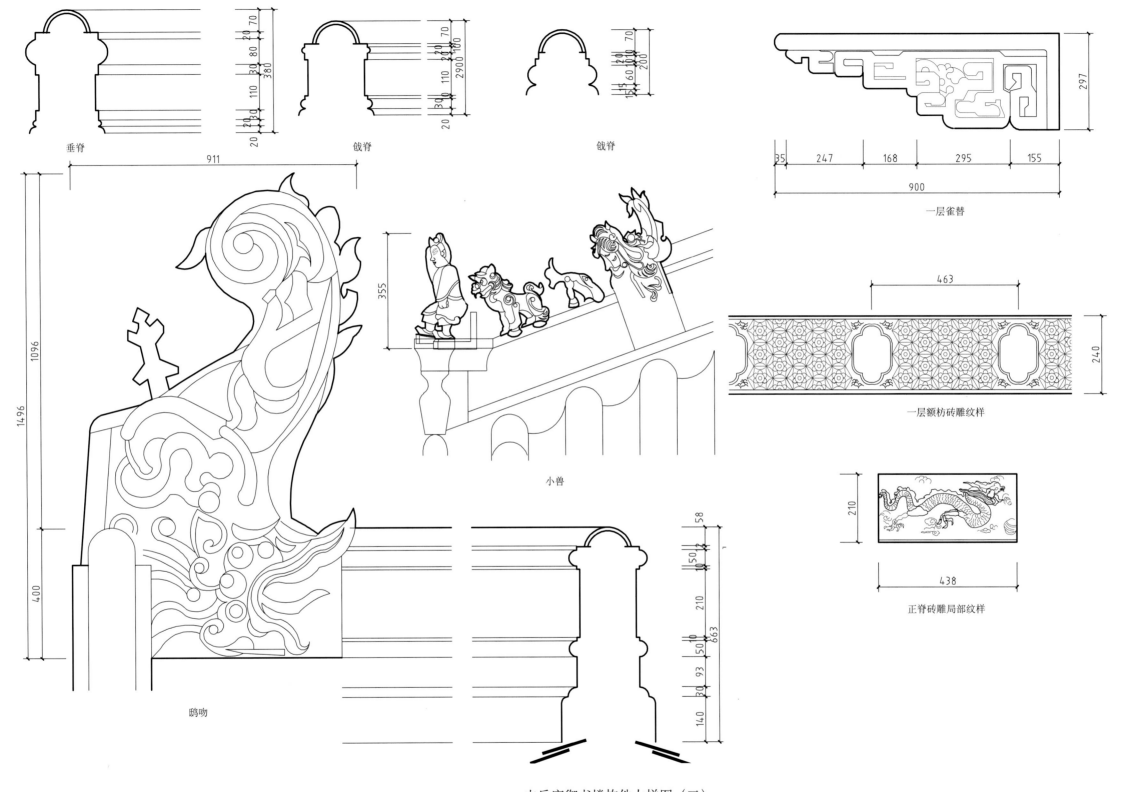

垂脊

戗脊

戗脊

一层雀替

小兽

一层额枋砖雕纹样

鸱吻

正脊砖雕局部纹样

中岳庙御书楼构件大样图（二）

Structural components of Yushulou of Zhongyue Temple Ⅱ

0   0.1   0.2m

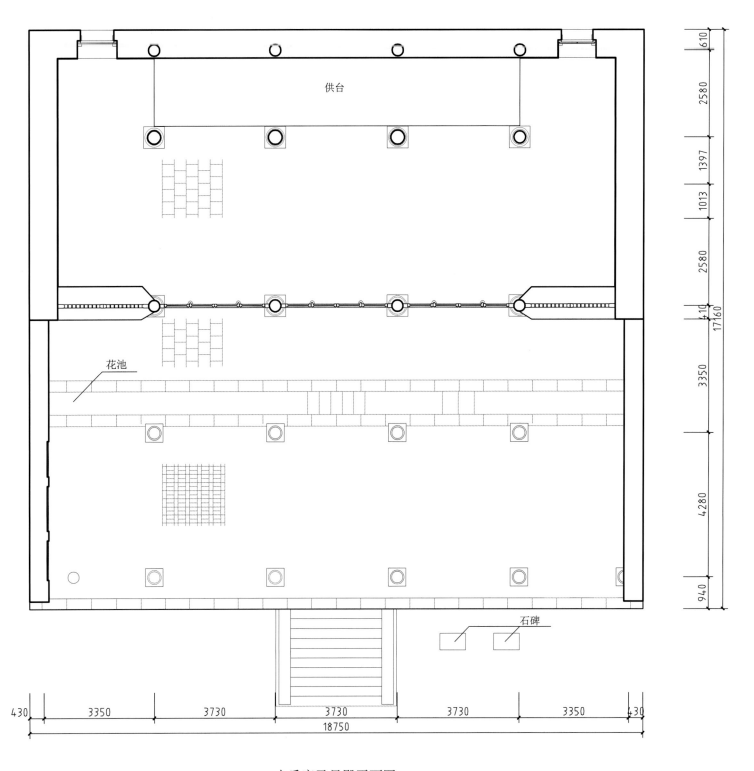

供台

花池

石碑

610
2580
1397
1013
2580
410
3350
4280
940
17160

430
3350
3730
3730
3730
3350
430
18750

中岳庙圣母殿平面图
Plan of Shengmudian of Zhongyue Temple

N

0 1 2m

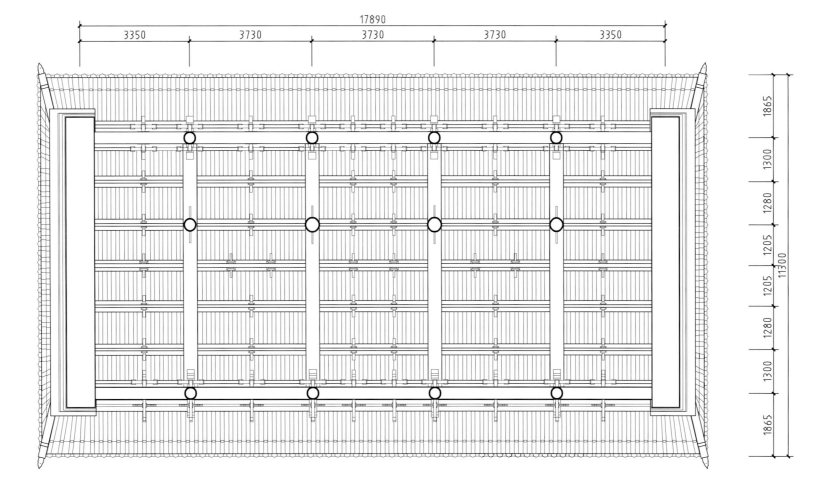

中岳庙圣母殿梁架仰视平面图
Framework plan of Shengmudian of Zhongyue Temple as seen from below

0 1 2m

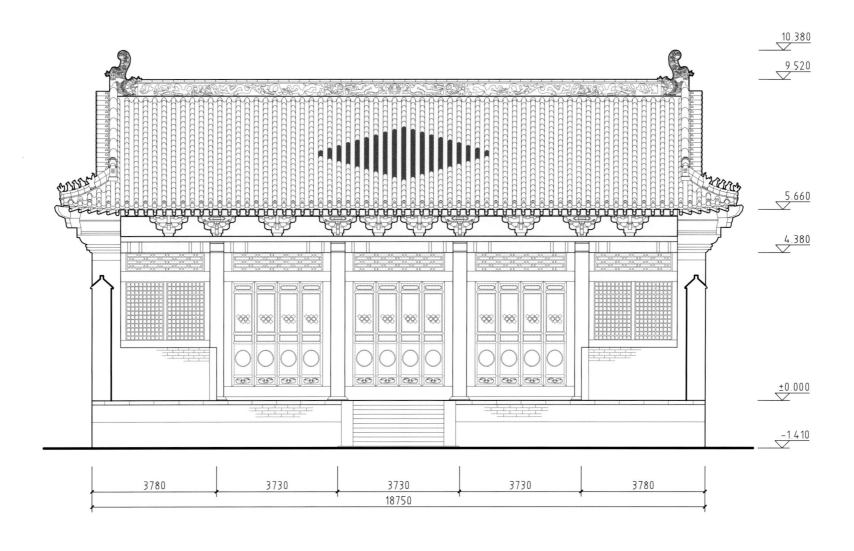

10 380

9 520

5 660

4 380

±0.000

-1.410

| 3780 | 3730 | 3730 | 3730 | 3780 |
18750

中岳庙圣母殿正立面图
Front elevation of Shengmudian of Zhongyue Temple

0 1 2m

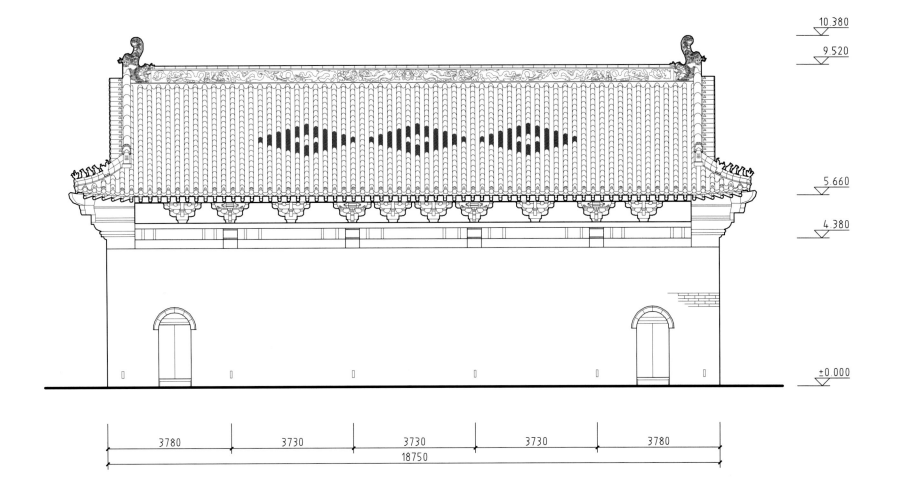

10 380

9 520

5 660

4 380

±0 000

3780　　3730　　3730　　3730　　3780

18750

中岳庙圣母殿背立面图

Rear elevation of Shengmudian of Zhongyue Temple

0　1　2m

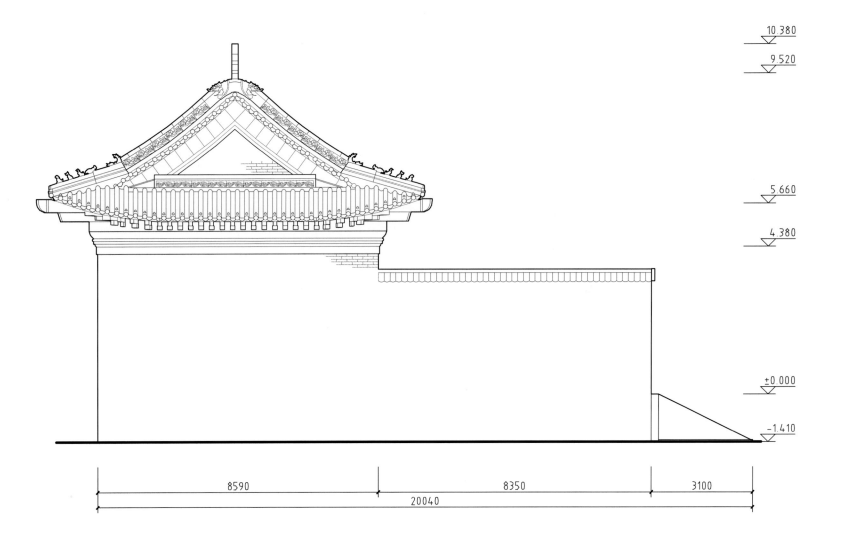

10.380

9.520

5.660

4.380

±0.000

-1.410

8590

8350

3100

20040

中岳庙圣母殿侧立面图
Side elevation of Shengmudian of Zhongyue Temple

0 1 2m

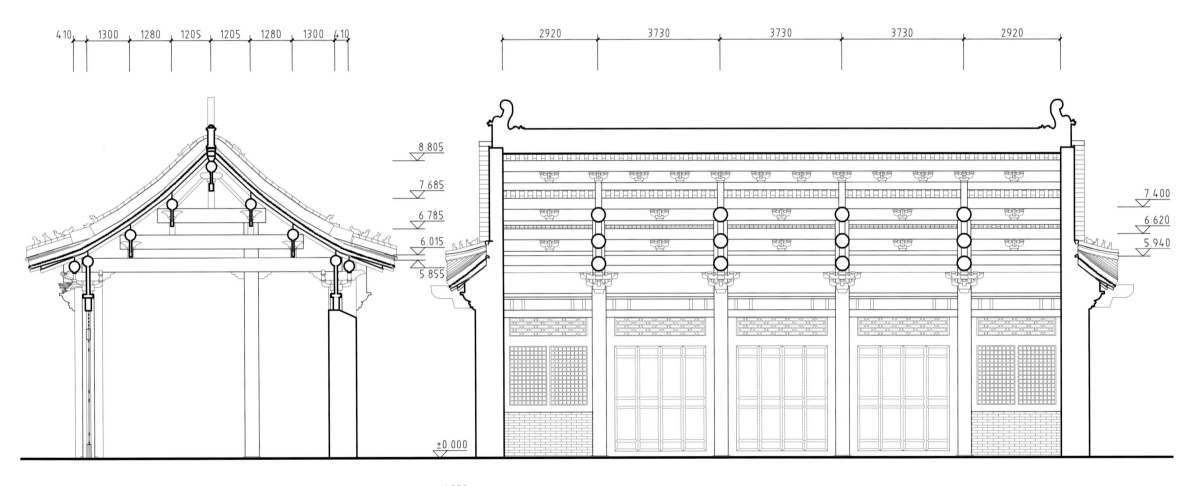

410 1300 1280 1205 1205 1280 1300 410

2920 3730 3730 3730 2920

8.805
7.685
6.785
6.015
5.855

7.400
6.620
5.940

±0.000

-1.350

中岳庙圣母殿横剖面图
Cross-section of Shengmudian of Zhongyue Temple

中岳庙圣母殿纵剖面图
Longitudinal section of Shengmudian of Zhongyue Temple

0 1 2m

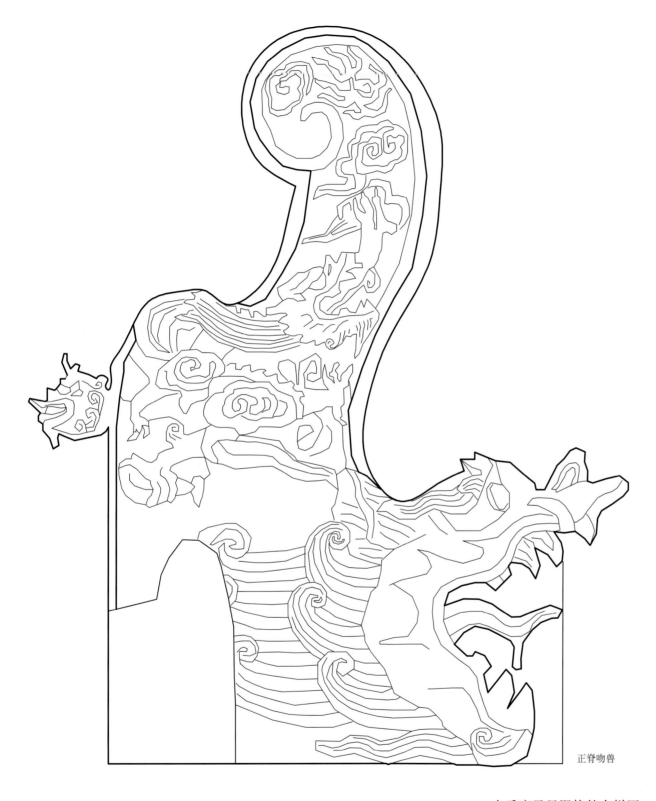

正脊吻兽

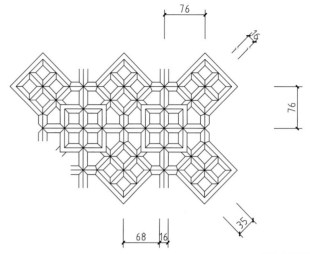

门上木格栅

门上木浮雕

中岳庙圣母殿构件大样图
Structural components of Shengmudian of Zhongyue Temple

0   0.1   0.2m

Name List of Participants Involved in Surveying and Related Works

Surreying and Mapping:

JIA Yaodong, SHAO Changkui, XIE Zhihan, XU Lianjun, HUANG Min, JI Wanjing, MA Jin, WANG Yue, SUN Li, ZHOU Danxi, LUO Jing, YI Lingjie, ZHANG Shuo, ZHANG Jingjing, LIU Xiaohe, SU Haixing, LI Rongxin, ZHANG Yuben, XIE Dan, LIN Tianpeng, PU Jieyu, ZHANG Yuanling, SHI Xiayao, JIN Xiangyu, LI Chunguang, WANG Yan, LI Qiang, YAN Jinbo, AI Hui, MENG Fanli, QIN Zhengdong, SHEN Xue, SUN Wenhao, WU Xiao, LUO Yu, Yoshiyama Junko, BAI Hao, ZHANG Rui, YONG Rong, PENG Lin, XIA Di, GU Shuwen, ZHOU Yajie, DONG Yao, LI Yu, LIU Yaosen, MA Li, DENG Yihong, ZHANG Guoqiang, LI Gang, ZHENG Yongya, CHEN Xiaojuan, SHEN Qinwen, BAI Dong, WAN Wenguang, QIN Xiao, ZHANG Dan, BO Le, LIU Li, GU Lina, TAN Yaning, HAO Shimeng, LI Yeguo, YANG Guang, QIAN Chengyuan, LIU Kun, LI Lu, LI Zhentao, WANG Ge, GONG Shijie, LIAO Huinong, LIU Chang, ZHENG Liang, XU Xiaoying, YANG Bo, DUAN Zhijun, XIN Huiyuan, JIANG Dongcheng, LI Dehua, HE Congrong, LI Luke

Translater in chief:

Alexandra Harrer

Team Members:

Alexandra Harrer, Michael Norton

参与测绘及相关工作的人员名单

测绘人员：

贾耀东　邵昶魁　谢智翰　徐连军　黄　敏　季婉婧　马　津
王　悦　孙　黎　邹丹曦　罗　晶　易灵洁　张　硕　章晶晶
吕晓荷　苏海星　李荣欣　章宇贡　解　丹　林天鹏　蒲洁宇
张元龄　史夏瑶　金翔宇　李春光　王　妍　李　强　闫晋波
艾　慧　孟凡理　沈　雪　孙文昊　吴　霄　罗　宇
吉山淳子　白　皓　张　蕊　戚征东　彭　琳　夏　荻　顾澍雯
周亚杰　董　瑶　李　煜　刘垚森　马　丽　邓一泓　张国强
李　港　郑泳雅　陈晓娟　沈勤雯　白　栋　万文光　秦　啸
张　丹　薄　乐　刘　利　谷丽娜　谭雅宁　郝石盟　李晔国
杨　光　钱程远　刘　坤　李　璐　李振涛　王　舸　龚树节
廖慧农　刘　畅　郑　亮　徐晓颖　杨　博　段智君　辛惠园
姜东成　李德华　贺从容　李路珂

英文统筹：

［奥］荷雅丽

英文翻译：

［奥］荷雅丽　Michael Norton

图书在版编目(CIP)数据

嵩山建筑群／清华大学建筑学院编写；王贵祥等主编.
—北京：中国建筑工业出版社，2019.6
（中国古建筑测绘大系·宗教建筑和礼制建筑）
ISBN 978-7-112-23463-9

Ⅰ.①嵩… Ⅱ.①王… Ⅲ.①嵩山－宗教建筑－建筑艺术－
图集 Ⅳ.①TU-885

中国版本图书馆CIP数据核字（2019）第047222号

丛书策划／王莉慧
责任编辑／李 鸽 陈海娇
英文审稿／［奥］荷雅丽（Alexandra Harrer）
责任校对／王 烨

中国古建筑测绘大系·宗教建筑与礼制建筑

嵩山建筑群

清华大学建筑学院 编写

王贵祥 刘 畅 廖慧农 贺从容 主编

*

中国建筑工业出版社出版、发行（北京海淀三里河路9号）
各地新华书店、建筑书店经销
北京方舟正佳图文设计有限公司制版
北京雅昌艺术印刷有限公司印刷

*

开本：787×1092毫米 横1／8 印张：32½ 字数：845千字
2020年6月第一版 2020年6月第一次印刷
定价：258.00元
ISBN 978-7-112-23463-9
（33774）